GOSSAMER THREADS

The Nearly Invisible and Nearly Unbreakable Chemistry that Connects Everything

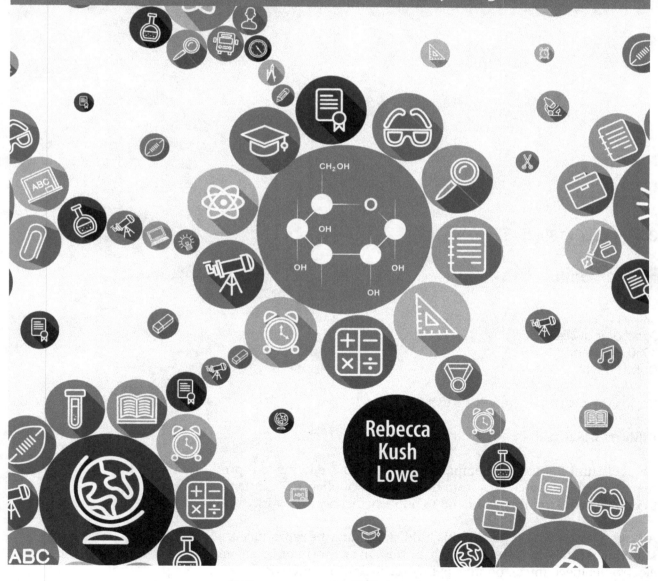

Rebecca
Kush
Lowe

Kendall Hunt
publishing company

Table of Contents

About the Author

Rebecca Kush Lowe teaches chemistry at Shippensburg University, and lives with her husband and family in Harrisburg, Pennsylvania.

Dedication

This book is dedicated to the men in my life: my husband Wayne, whose support brought this book to fruition; my sons Josh, Andrew, Calvin, and Sam, who bring me joy and help me to see the world with fresh eyes each day; and my father, the late William J. Kush, who shared with me his love of chemistry and his love of life. Thanks, guys. I love you.

Introduction

What is chemistry? Is it going to be boring?

What kind of things do you think of when you hear the word *chemistry*? Maybe you have memories of high school chemistry, where you memorized the periodic table and did lots of conversion problems. Or maybe you have images of people in lab coats and goggles swirling colored liquids around in funny-shaped flasks. Chemistry is so much bigger than these visions. Chemistry is *the study of matter and its interactions*. And chemistry is not boring, or limited to a laboratory or special clothing and equipment. Chemistry is the basis for life, it has been at the root of human civilization, and it is the driving force behind the development of new materials and technologies. It's also the reason why you like some foods and not others. Chemistry is the substance of all of our modern (and ancient!) medicines. Chemistry holds the keys to understanding the past, maximizing the present, and dreaming into the future. Chemistry is awesome!

So, if chemistry is the study of matter and its interactions, it's appropriate to know what *is matter*. All things in our universe are made up of matter and/or energy. Anything that has *mass* and *volume* – takes up 3-D space – is defined as being matter. Almost everything you come into contact with on a daily basis is matter. Even you are matter! And all matter, even you, are made up of the same building blocks in this universe. These building blocks are *atoms*. They are the basic component of all matter, and they themselves are all composed of the same pieces: protons, neutrons, and electrons. More detail will be given later about atoms, but for now, understand that protons and neutrons are the parts of an atoms that make up the lion's share of that *mass* we said characterizes matter. Electrons, which have a much smaller mass than protons and neutrons, define the majority of the *volume* of atoms and of matter. So we need all of these pieces to make up all the matter in our universe.

What tools help to understand chemistry? How can I be successful in chemistry?

When we want to look at interactions of matter, it's good to have some tools along to help. The most recognizable tool of any chemist is the periodic table. It looks so orderly and reasonable – look at how the numbers above each symbol increase so smoothly starting at the upper left of the table in the box with H in it. And yet there is variety in the table, too, among the printed symbols and names that represent different types of matter – called *elements*. Each box in the table represents information about a unique kind of matter that we have discovered in our universe or created here on Earth. We will use the periodic table as a source of information about the matter we come into contact with.

One of the first things we can learn from the periodic table is that it tells us that there are two general types of matter in our universe: *metals* and *nonmetals*. There is a bold zigzag line starting below boron (B) and above aluminum (Al), which proceeds down and to the right among other elements of the table. This zigzag line tells us at a glance that all elements that are arranged to the left and below that zigzag line would be classified as metals. You can see that this makes most of the elements in our inverse 'metals'. All the elements arranged to the right and above of the zigzag stairstep line are considered *nonmetals*. There are considerably fewer of these kinds of matter; however, this little group has a huge role in the chemistry we encounter on a daily basis and in our own bodies.

What does it mean to say that an element is a *metal* or that a type of matter is *metallic*? It means that it looks and acts consistent with the way we understand metals to look and act. A metal is usually solid at room temperature, shiny, able to carry electrical current, able to transfer heat, able to be reshaped (malleable), and able to be pulled into a thin wire (ductile). Examples of metals include sodium (Na), iron (Fe), tungsten (W), lead (Pb) and copper (Cu). Nonmetal elements are also classified according to their characteristics. Nonmetals are sometimes solids are room temperature, like carbon (C), but many are gases. Nonmetals are not good conductors of heat or electricity and have very limited abilities to be manipulated into different shapes. Other examples of nonmetals include boron (B), neon (Ne), sulfur (S), and argon (Ar).

The periodic table is the go-to source for information about the matter we study. It has even more secrets to share, which we will reveal as we go along. For now, you already know that if you see an elemental symbol, you can go to the periodic table and find the name of the element that goes with that symbol, and you can state with conviction whether that element is a metal or a nonmetal. You are already securing your reputation as the most interesting person at the dinner table!

What other tools to chemists use to maximize what they learn about and can do with matter? Perhaps the most important tool, maybe even more important even than the periodic table, is their sense of curiosity. *What is that stuff*? That is so cool! *What makes that material act that way? How can this matter be used to teach me about the universe? How can what I have learned help me create new materials and products?* This last question brings another tool with it: ingenuity. Modern materials like plastics have grown out of a desire to use knowledge about existing materials to create new materials that have qualities better suited to modern needs. One surprising member of the scientific toolbox is *serendipity* – a chance discovery that turns out to be useful or beneficial in some way. We will be identifying serendipitous materials as we come across them in our study of chemistry. Finally, all chemists have perseverance in their toolbox. They keep working to understand concepts, they try and try again after failure, and they push forward, trusting that the wonder of chemistry will not disappoint. Chemists are a tenacious bunch! While you read this book, you are a chemist. Welcome to the adventure!

Branches of Chemistry

Organic Chemistry

Chemistry as a discipline is like a huge tree that has two main branches. These branches are called *organic chemistry* and *inorganic chemistry*. Much of the material we will study grows on the organic chemistry portion of this massive subject tree. What does it mean to be 'organic'? The oldest versions of this word reflect an understanding that material can be obtained out of living things. Therefore, the original use of the word organic meant a substance that is obtained from a living organism. Fluids like blood or urine were called 'organic'. This definition is a good start for today's chemists.

People who are not chemists (that's most of us!) think of the word 'organic' with a different meaning. Usually it means something that is fresh, grown without pesticides or other added chemicals, and is considered healthy, or at least healthier than something that does not have the word 'organic' on its label. This kind of interpretation is much more rooted in culture and business than in science. This kind of usage is broad and has many different nuances depending on the context in which it is being used.

Today's chemists have a more specific meaning of the word 'organic'. Organic reflects the chemical composition of a material. Organic chemistry centers around matter which has carbon (C) as it's main structural component in the material. You can see that carbon is a nonmetal. What's also true about organic chemicals is that they most often partner carbon with other nonmetals from the same region of the periodic table. Carbon partners easily with hydrogen (H), oxygen (O), and nitrogen(N). We also will see materials that include sulfur (S) and phosphorous (P), or even occasionally fluorine (F), chlorine (Cl), and bromine (Br). We will use this specific definition of organic throughout this book.

What kinds of organic chemicals exist? There are millions of known organic compounds. Some examples are DNA – deoxyribonucleic acid – the genetic material we each carry. The structure of each person's DNA is a beautiful scaffold of carbon, hydrogen, oxygen, nitrogen, and phosphorous. The chemical chlorophyll that enables green plants to produce food and oxygen from sunlight is an organic compound. Blood is a mixture of amazing organic materials like hemoglobin, which includes a partnership of nonmetal organic structures with metal atoms of iron. Organic chemistry underpins all of life here on earth.

Inorganic Chemistry

Inorganic matter usually contains metal elements, or is mainly composed of nonmetals that are not carbon. There are also many varieties of inorganic materials, and certainly there is more variety in the chemical composition among inorganic materials. Just look at how many metallic elements there are! Examples of inorganic materials include aluminum foil, which is composed of (ahem!) aluminum (Al). Or table salt, which contains sodium (Na) and chlorine atoms. What about that beautiful opal birthstone? It has silicon and oxygen atoms. Basically, if a material does not contain carbon as the main structural component, it's classified as inorganic. Organic chemistry gets a lot of glory, being the chemistry responsible for, like, *life*, but inorganic chemistry is responsible for much of the chemistry of the earth's formations, the composition of the earth's atmosphere, and for the forces that shaped our early universe.

Expressing Chemistry

Elements

The periodic table catalogs the unique types of matter that are known to exist on earth and in the universe. We have already been talking about the elements as those types of matter represented by unique names and symbols organized in increasing numerical order on the periodic table of the elements. Elements are the building blocks of chemistry, in the same way that letters are the building blocks of words. Elemental symbols and names provide a way to share information about matter through writing. Elemental symbols and names represent real matter, and allow scientists to present and to propose the interactions of matter without having to go into the lab or show a demonstration every time they want to review how the chemistry is happening.

Compound

Not all particles of elements exist alone. Most matter on the earth is some kind of chemical combination of two or more elements. These are defined as *compounds*. Compounds are represented by elemental symbols that are grouped together. For example, table salt is known to be composed of sodium (Na) and chlorine (Cl). When a chemist talks about table salt, it is written NaCl, representing the chemical combination of sodium and chlorine. According to the periodic table, sodium is a (solid) metal and chlorine is a (yellowish gas) nonmetal at room temperature. Both of these elements are poisonous by themselves. But table salt is a white crystalline solid at room temperature that we need to survive. So this set of facts reinforces the understanding that compounds have properties which are unique from the properties of the elements which make up the compounds. When chemical interactions – transformations – take place, those interactions result in changes to the matter which also changes the physical and chemical properties which define that matter.

The term 'compound' specifically refers to a chemical connection between two or more atoms of different identities. But often the word 'molecule' is also used, but there is a subtle difference in the understanding of this word. The word 'molecule' is thought of as more general than 'compound' because it simply implies a substance with more than one atom in it. There is no specification in the word 'molecule' about the types of atoms in the substance. Molecules are also understood to be the physical representation of groups of atoms, many of which are compounds, too.

How can chemists communicate about elements, atoms, compounds, and molecules?

Because atoms and molecules are often way too small to be seen with even some of the most powerful microscopes, chemists have a variety of ways to communicate what atoms and molecules look like and act like. Models are a really great way to share an atom or molecule. Models show three-dimensional structure and the relationship between atoms in a molecule. For example, consider the substance **glucose**, which is the material our bodies use as fuel to do all the things needed to survive. This is a picture of a model of glucose:

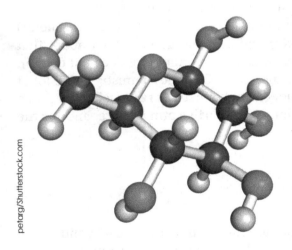

petarg/Shutterstock.com

In this model, the black balls represent carbon, the red balls represent oxygen, and the white balls represent hydrogen. If you look carefully at the model, you should be able to see that there are 6 black balls, 6 red balls, and 12 white ones. This is meant to indicate that in a real molecule of glucose, there are 6 atoms of carbon, 12 atoms of hydrogen, and 6 atoms of oxygen, arranged relative to one another the way the balls are arranged in the model. The straight rods connecting the balls to each other in the model represent the chemical connection the atoms have to each other. So we can understand from this model what atoms of the given elemental types are present, how they are connected, and how the entire group arranges itself in space. This is really great information! But sometimes models are not practical. Carrying around a model kit can be cumbersome, and it takes time to build a molecule, especially some of the really big ones like proteins and polymers.

What other options exist to share information about the composition of compounds?

The simplest and most concise way to talk about compounds is to use a *condensed chemical formula*. This is chemical 'word' that gives the elemental composition of any given compound, without any information about structure. It's kind of the opposite extreme to a molecular model. In the model above, we saw the compound's composition in the picture. To represent this same compound with a condensed chemical formula, we would write $C_6H_{12}O_6$. Can you see the relationship between the molecular model and it's condensed chemical formula? Each type of atom, represented by a colored ball in the model, is instead written with its *elemental symbol*. This is the point where having a periodic table comes in handy. Carbon is represented by the capital letter C, hydrogen is capital H, and oxygen is capital O. Then each elemental letter is followed by a number, written as a subscript, that tells *how many of each atom of that type* exist in the compound. If you see a condensed chemical formula and there is no subscript following an elemental symbol, this means there is only one of that kind of atom in the compound. Consider the following condensed chemical formulas and identify the elemental composition of each one:

Compound	Elemental composition
$C_6H_{12}O_6$	6 Carbon atoms 12 Hydrogen atoms 6 Oxygen atoms
C_2H_6O	2 Carbon atoms 6 Hydrogen atoms 1 Oxygen atom
NaCl	1 Sodium atom 1 Chlorine atom
H_2O	2 Hydrogen atoms 1 Oxygen atom
H_2SO_4	2 Hydrogen atoms 1 Sulfur atom 1 Oxygen atom

Is there something in between these two extremes? What if we want some information about structure, but still don't want to build a model? There are options. Two of the main options include structural formulas and skeletal formulas.

Structural formulas show all the elements in a compound and show the relative spatial relationship of each atom of each element in the compound, but in a two-dimensional way. For example, here is a structural formula representing our compound glucose:

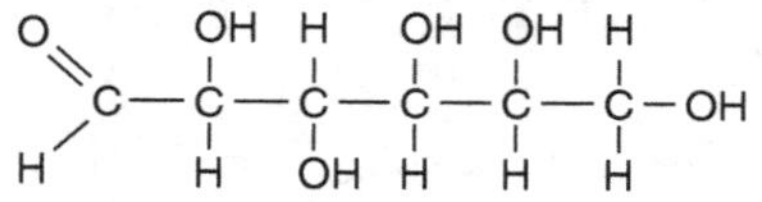

Looking closely, you should see that the structural model is true to both the molecular model and the condensed chemical formula. This is very important. Formulas must accurately reflect the compounds they represent. The structural formula allows the reader to see all the types of elements in the compound as well as have a basic understanding of the relative position each atoms of each element has to others in the compound. Structural formulas are great at communicating composition and structure, more completely that condensed chemical formulas do, but are more practical than building a model every time you want to discuss a compound.

Many of the compounds we will look at in this course are organic, like glucose. You have already learned that organic chemistry is identified mainly by its use of carbon as a main structural component. Looking at glucose, an organic compound, this is true as well. Carbon atoms form the central backbone of the compound, and other atoms branch off from them. When discussing organic compounds, there is another type of formula that can be used to discuss structure and composition, but it eliminates the need to use the symbol for carbon and streamlines the use of the symbol for hydrogen as well. This is called the *skeletal formula*. Here is the skeletal formula for glucose:

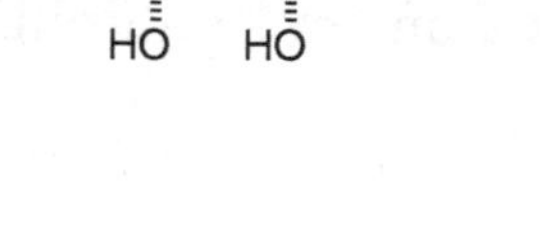

What has changed between the structural and skeletal formulas for glucose? The symbol for carbon has been eliminated, but it has not been removed from the compound itself, just from the skeletal formula. Each time a carbon is represented, there is a meeting of two lines at an angle, like a zigzag backbone. This is actually a nod to the actual structure, if you look back at the molecular model for glucose. Those black balls didn't lie in a straight line, but in reality are at fixed angles to one another. In addition, hydrogen atoms that are bonded directly to carbon atoms are not written directly, but are assumed to be present. Solid lines represent spatial relationships that lie in the same plane. Dotted lines represent spatial orientation going away from you. The solid black triangle represents a spatial orientation leaning out toward you. These conventions once again attempt to bring more three-dimensional truth to a two-dimensional representation. Not all skeletal formulas use these specific conventions. But make no mistake: skeletal formulas must still represent the truth about the composition of a compound. An organic chemist (such as you are at this moment) would understand that carbons are present at each juncture of straight lines, and that hydrogen atoms attached directly to carbons won't be written down, but are still there.

How can skeletal formulas be translated back into structural and condensed chemical formulas?

One key to this feat is to know that carbon always has four chemical relationships – four 'bonds' – represented by straight lines to other atoms in a formula. The reasons for this reality will be discussed later on. But in order to discuss the identity of a compound with others, it is helpful to compare the different kinds of structures. Consider this skeletal structure, representing the main ingredient in the common kitchen substance, vinegar:

Which would also have the structural formula of

Acetaldehyde

See how the symbols for carbon have been reinserted at the places where the central backbone of the skeletal formula bent and ended? The general rule is that at every end of a line and every bend in a line, a carbon exists, unless otherwise indicated. And once the carbon atoms are placed in the structure, we look to see how many relationships those carbons have. Each line leading out of where a carbon is represents another chemical relationship to that carbon. These chemical relationships are usually called *bonds*. Bonds to all other elements are shown as straight lines in a skeletal formula, except bonds directly between carbon and hydrogen. If a skeletal formula does not show 4 bonds coming from any given carbon, you can assume that any missing bonds are to a hydrogen. The structural formula reinserts these assumed bonds.

Once all the elements are represented, a condensed chemical formula can be created by counting up how many of each type of atom are present and summarizing them in the condensed chemical formula. In this example, the condensed chemical formula for our ingredient – also called acetic acid – is $C_2H_4O_2$.

Each type of chemical formula has its advantages and disadvantages. Some are concise, like a condensed chemical formula. Some reveal more structure, like structural and skeletal formulas. Some allow structure to be more quickly understood using common assumptions, as the skeletal formula does. Some situations benefit more from the use of one kind over another. You should be able to decipher the information each one provides to you.

Chemical Reactions

Chemists name chemical interactions as 'chemical reactions'. Chemical reactions are observed to have happened when certain physical and chemical evidence presents itself. A chemical reaction is said to have occurred if 1) a color change in the matter is observed; 2) a gas is created within the chemical system (look for bubbles to form); 3) light is given off by the chemical system (chemiluminescence); 4) the temperature of the chemical system changes – meaning that heat is being created (an exothermic reaction) or heat is being absorbed (an endothermic reaction); and/or 5) a solid (called the 'precipitate') is created. Deeper study of chemical reactions reveals that all chemical reactions have as their primary driving force the movement of electrons between particles of elements and compounds. Changing the arrangement of electrons in an atom creates the material interactions that result in chemistry. More on this later.

Chemical Equations

Chemical reactions are amazing, but they aren't always convenient to bring along to show someone. So chemists use symbols to represent the chemical reaction. Writing a chemical reaction in symbolic form is the chemical reaction equation, or sometimes just the chemical equation. Chemical equations are certainly more convenient in some cases, but it's also true that the experience of the chemical reaction itself is lost. So there are some limitations to a chemical equation. On the other hand, a chemical equation shows very clearly the specific composition of each type of matter involved in the reaction as well as the relative amounts of each substance in the reaction. A chemical equation is like the written recipe for a chemical reaction which is like the ingredients that transform to become a delicious finished food product.

Consider this recipe for chocolate chip cookie bars:

Ingredients:

- 1 cup (2 sticks) unsalted butter
- ¾ cup packed brown sugar
- ¾ cup granulated sugar
- 2 eggs
- 1 tsp vanilla extract
- 2 ¼ cups flour
- 1 tsp baking soda
- 1 tsp salt
- 12 oz. chocolate chips
- 1 cup chopped walnuts

Directions: Preheat oven to 375°F. Cream butter and sugars. Add eggs, one at a time, then vanilla and mix until well blended. Stir together flour, baking soda, and salt, and add gradually to butter mixture until well blended. Stir in chocolate chips and nuts. Spread in an ungreased 9″ × 15″ jelly roll pan. Bake for 20–25 minutes or entire cake is evenly golden brown and a toothpick inserted in the center of the cake comes out with just a few crumbs clinging to it. Allow to cool in the pan, for 10 minutes, then cut into bars.

Yield: 48 bars

If we could write this recipe in chemical style, it would look something like this:

1 cup   1 ½ cups 2   1 tsp   2 ¼ cup   1 tsp   1 tsp   12 oz.   1 cup   48

Butter + sugar + eggs + vanilla + flour + baking soda + salt + chocolate chips +
walnuts → chocolate chip cookie bars

Notice that there are no directions for the cooking of the recipe written in this recipe 'equation', though the identity and relative amounts of each of the ingredients and the products are listed. In the same way, a chemical equation lists the ingredients – the reactants – and the products as well as the relative amounts of each.

Here is a reaction equation that represents a real chemical reaction performed in every green plant:

Photosynthesis reaction equation

In words:

carbon dioxide gas + liquid water → solid sugar + oxygen gas

In chemical symbols:

$$6CO_2(g) \quad + \quad 6H_2O(liq.) \rightarrow C_6H_{12}O_6(s) + 6O_2(g)$$

The chemical reaction equation shows the 'recipe' for photosynthesis. Like a cooking recipe, the 'ingredients' are listed on the left side of the arrow. These ingredients, called *reactants* in chemistry, have specific compositions – the chemical formulas for the compounds – and are used in specific amounts. Remember that within formulas, any numbers are subscripts and represent the number of atoms of each element in the compound being represented by any given formula. The large number to the left of any chemical formula, called a coefficient, dictates how many of that kind of molecule is needed to make the reaction proceed properly. The formulas on the right side of the arrow are the *products* of the chemical recipe. The reaction equation reveals how much of each product should be made by the reaction. If a coefficient is not present in front of a chemical formula, or a subscript value is not present after an elemental symbol within a formula, the number '1' is assumed.

So as we examine the equation for the reaction known as photosynthesis, we can see that 6 molecules of carbon dioxide – 6 CO_2 – will chemically combine with 6 molecules of water – 6 H_2O – and will result in the reorganization of the atoms into 1 molecule of a simple sugar – (1) $C_6H_{12}O_6$ – and 6 molecules of elemental oxygen – 6 O_2.

Notice also that there is a material balance that is obeyed in this chemical equation. Meaning that the number of atoms of each type on the reactant side are also represented in the same overall quantities on the product side, albeit in new arrangements. In fact, this obeisance is present in every chemical reaction in our universe. This is understood as *The Law of Conservation of Matter*. In other words, all the matter in the universe is already present, and no matter can be created anew nor destroyed permanently. This idea can be represented in broad terms by thinking about a seesaw beneath every chemical equation. This seesaw will teeter up on one side and down on another with the fulcrum of the seesaw under the arrow that separates the reactant and product sides of the equation. Imbalance will reign on the seesaw until the formulas and their coefficients support a balanced number of each type of element's atoms on each side of the arrow in the middle.

$6CO_2(g) + 6H_2O(liq.) \rightarrow C_6H_{12}O_6(s) + 6O_2(g)$		
Δ		
Element (symbol)	Number of atoms of element on reactant side of equation	Number of atoms of element on product side of equation
Carbon (C)	6 (from 6 molecules of CO_2)	6 (in 1 molecule of $C_6H_{12}O_6$
Oxygen (O)	18 (12 from 6 molecules of CO_2 and 6 from 6 molecules of H_2O)	18 (6 in 1 molecule of $C_6H_{12}O_6$ and 12 from 6 molecules of O_2)
Hydrogen (H)	12 (from 6 molecules of H_2O)	12 (in 1 molecule of $C_6H_{12}O_6$)

With this idea in mind, any chemical equation can be judged for its level of balance in this molecular seesaw.

Are these chemical equations balanced?

Fermentation: sugar → alcohol + carbon dioxide

$$C_6H_{12}O_6 \rightarrow C_2H_6O + 2CO_2$$

Element	Number of atoms on reactant side	Number of atoms on product side
Carbon (C)	6	4
Hydrogen (H)	12	6
Oxygen (O)	6	5

Conclusion: No, this equation is not balanced. The number of atoms of each element is not represented in the same amounts on both sides of the equation.

Combustion of propane: propane + oxygen → water + carbon dioxide

$$C_3H_8 + 5O_2 \rightarrow 4H_2O + 3CO_2$$

Element	Number of atoms on reactant side	Number of atoms on product side
Carbon (C)	6	12
Hydrogen (H)	12	6
Oxygen (O)	6	5

Conclusion: Yes, this equation is balanced. The number of atoms of each type of element is represented equally on both the reactant and product sides of the equation.

Action of an antacid (Neutralization):

stomach acid + sodium bicarbonate → salt + carbon dioxide + water

$$HCl + NaHCO_3 \rightarrow NaCl + CO_2 + H_2O$$

Element	Number of atoms on reactant side	Number of atoms on product side
Hydrogen (H)	2 (1 from 1 molecule of HCl and 1 from 1 molecule of $NaHCO_3$)	2
Chlorine (Cl)	1	1
Sodium (Na)	1	1
Carbon (C)	1	1
Oxygen (O)	3	3 (2 in 1 molecule of CO_2 and 1 in 1 molecule of H_2O)

Carbon is unique, but all atoms are the same?

Carbon is unique among the elements of the periodic table because of its ability to easily bond with – in this case, *to bond* means an ability to share electrons easily with – other atoms of carbon as well as many other nonmental elements. This ability to partner makes the variety of organic ('carbon-based') matter on earth nearly limitless in the kinds of combinations that can be formed.

But there is one rule that carbon atoms **must** obey. Carbon atoms can only have 4 bonds – 4 electronic partnerships – at a time. No more, and no less, in stable organic compounds. Why is this? Because of the way carbon atoms are built. Remember that all atoms have the same building blocks of protons, neutrons, and electrons. The protons and neutrons in every atom (except for hydrogen, which only has 1 protons and no neutrons) exist in the physical center of the atom and form a dense but very massive nucleus of an atom.

The proton is understood as being massive, on the atomic scale, and as carrying a charge – an electromagnetic signature of sorts – which is defined as 'positive'. Neutrons are the same size as protons, and as massive, but don't have any kind of charge; neutrons are 'neutral', as if they are the Switzerland of the atomic world. Electrons are an interesting partner to these other players. Electrons are incredibly light, especially compared to protons and neutrons, but carry a charge equal in strength but opposite in signature to the proton. Electrons are understood as carrying a 'negative' charge, and they don't exist in the densely packed, massive nucleus, but instead exist in regions of space around the nucleus, which most scientists today call the 'electron cloud'. These realities are crucial to our understanding of chemistry, physics, and biology, though what we take for granted as understanding today was hundreds of years in the making by scientists all around the world.

Subatomic particle	Atomic location	Mass (kg)[1]	Charge
Proton	Nucleus	1.67262×10^{-27}	+1
Neutron	Nucleus	1.67493×10^{-27}	0
Electron	Electron cloud	9.109×10^{-31}	−1

With this understanding in place, we can also pull back and look at the bigger picture as well as the atomic one. The periodic table also has information about the atomic makeup of any atom of any given element. Any one element's region on the periodic table is a box with numbers and letters in

---

[1] *Encyclopædia Britannica Online,* s. v. "proton", accessed February 16, 2015, http://www.britannica.com/EBchecked/topic/480330/proton.

it. These numbers and letters are going to help us understand why carbon likes to have 4 bonds, as well as help us predict how other elements choose to bond.

'Why, it's elementary, my dear Watson!'

It makes sense to use carbon as our example. Look at carbon's place on the periodic table. It is a nonmetal element as it lies to the right of that stair-step line that starts just below carbon's left-hand neighbor, boron (B). The box in which carbon's symbol lives has the name of the element below the symbol and also has two numeric values. The number printed above the symbol is a whole number, called the *atomic number*. Carbon's atomic number is 6. It tells us that carbon is the 6th element on the periodic table, but it also reveals more than that.

The atomic number of an element tells us how many protons exist in the nucleus of an atom of that elemental type. Carbon's atomic number, 6, also reveals that every carbon atom contains 6 protons in its nucleus. This is important because the number of protons in an atom dictates that atom's material identity. All carbon atoms contain 6 protons, no more and no less, in the nucleus. The truth of the atomic number is that every atom is unique in the number of protons it contains, which partially accounts for the unique properties of the atoms, whether single and elemental, or combined into compounds.

Therefore, ask yourself these questions: how many protons are in an atom of hydrogen (symbol H)? The atomic number of hydrogen is 1; therefore, the number of protons in any hydrogen atom and all hydrogen atoms is 1. What about lead (Pb)? The atomic number of lead is 82; so every lead atom contains 82 protons. Pretty straightforward, right?

There is more to discover about atoms using the periodic table. The atomic number also reveals the number of electrons that an elemental atom, existing alone, would carry in its electron cloud. So, in a carbon atom, 6 electrons are present in the cloud around the nucleus. How many electrons in an atom of hydrogen? (1) How many in an atom of lead? (82). Unlike protons, though, the number of electrons that an atom *can have* is not fixed like numbers of protons are. Electrons are in constant motion, and chemistry happens when they move so far as to leave one atom and move to another, even if it's temporary. Electrons can be moved more permanently, from one atom to another, in a concept called *ionization*, which we'll talk more about later.

What can you say so far about a carbon atom? It is a nonmetal atom. It always contains 6 protons, which exist in the nucleus of the atom and which compose a significant portion of the mass of the carbon atom. A carbon atom which is not in a compound will have 6 electrons, which have enough negative charge among them to balance the positive charge contributed by the much bigger protons. It's not a mistake that atoms have the same number of protons and electrons when in their pure, elemental states.

But atoms have another component, neutrons. Can we use the periodic table to reveal the secrets of the neutron? Absolutely. But it's not in the atomic number. It's in the *atomic mass*, the numeric value below the elemental symbol and name. This number is not a whole number, but usually contains a decimal portion as well. The atomic mass reflects the variety that exists even among atoms of the same element that are known to exist in the universe. That variety exists in a population is not an unfamiliar concept. Any group of similar items will have some differences among its members. For an atom, this continuum of variety results in atomic masses that differ even among the members of a population of atoms of the same elemental identity.

Encyclopædia Britannica Online, s. v. "neutron", accessed February 16, 2015, http://www.britannica.com/EBchecked/topic/410919/neutron.

Encyclopædia Britannica Online, s. v. "electron", accessed February 16, 2015, http://www.britannica.com/EBchecked/topic/183374/electron.

If all atoms of an element are supposed to be the same – they all have the same atomic number, so they are all identified as being the same element, after all – what causes the variance in masses? The number of neutrons in an atom can vary a little among some members of a population of atoms of the same type. Neutrons neither dictate an atom's identity (only protons can do that) nor control the chemistry that governs most reactions (that's the electrons' job); but they do provide mass to an atom. The atomic mass number helps us determine the average number of neutrons we can expect most atoms of any given element to have. How? Some simple subtraction. Here's how it works:

1. Find the atomic mass number and round it to the nearest whole number. For carbon, the atomic mass is 12.011 atomic mass units (amu). Rounded to the nearest whole number, this means '12'.

This rounded number represents the average mass of an atom of that type. As the mass of an atom is primarily concentrated in its protons and neutrons in the nucleus, this means that a rounded mass of '12' is due to the general mass units of the protons and neutrons of the carbon atom. Each proton and neutron is defined as having approximately 1 mass unit each. But, as we already know how many protons the carbon atom had. . .

2. Subtract the number of protons (found using the atomic number) from the atomic mass (representing the total protons and neutrons in the atom) to find the number of neutrons in any atom. Generally speaking, this looks like:

Atomic mass minus atomic number = Number of neutrons
(Number of protons and neutrons) – (Number of protons) = Number of neutrons

And for carbon specifically, we have:

Atomic mass 12 – atomic number 6 = 6 neutrons

Therefore, we have all the information we need to understand what any carbon atom brings to a chemical reaction: 6 protons, 6 neutrons, and 6 electrons. This reality informs how that same carbon atom will choose to interact when given the opportunity to do so chemically.

So, back to the original question: What makes carbon so unique in its ability to bind chemically with other carbons atoms so well? This answer can be answered a little more by looking again at where carbon sits in the arrangement of the periodic table. Carbon sits in the second row, in the fourth position starting from the left end of the row beginning with lithium (Li). There are four more elements that sit to the right of carbon. The number of elements listed in a horizontal row in the periodic table also reflects how many electrons there is space for in the electron cloud of the element s of that row. This is especially true for nonmetals like carbon. So, looking at this row where carbon lives, how many elements are there in that row on the periodic table? There are eight elements. This reflects the reality that elements which fall in this period of the table can hold as many as 8 electrons in the outermost region of the electron cloud, the region responsible for bonding. How many electrons does carbon have already? According to the atomic number, a carbon atom has 6 electrons, but according to the periodic table, there are 4 in the outermost region, and room for 4 more. These four more spaces for electrons are how new bonds will be formed. The number of bonds available to be formed is equal to the number of spaces available in the electron cloud. This is especially true for the nonmetal elements in the first two periods of the table, and this region is largely responsible for the majority of compounds found in living systems.

Let's test what we know so far. Complete the following table in your notebook, using your periodic table.

Element (symbol)	Period (horizontal row)	Number of elements in the period	Number of protons in an atom	Number of neutrons in an atom	Number of electrons in an atom	Number of bonds likely to be formed
Hydrogen (H)	1	2	1	0	1	1
Helium (He)	1	2	2	2	2	0
Carbon (C)	2	8	6	6	6	4
Nitrogen (N)	2	8	7	7	7	3
Oxygen (O)	2	8	8	8	8	2
Fluorine (F)	2	8	9	10	9	1
Neon (Ne)	2	8	10	10	10	0
Chlorine (Cl)	3	8	17	18	17	1
Argon (Ar)	3	8	18	22	18	0

We now have a slightly better understanding of how atomic structure determines chemical behavior in nonmetals, and of the relationship between the arrangement of the periodic table and atomic structure of those elements. With this in mind, we can now better understand and predict how an organic compound will come together and how it will behave. This is crucial to grasping the realities of the chemistry that surrounds us.

FOOD CHEMISTRY

UNIT 1

'What are little girls made of? Sugar and spice and everything nice.
That's what little girls are made of'.

-part of a 19th-century nursery rhyme

SUGARS AND ARTIFICIAL SWEETENERS

Sugars are delicious. Human biology is designed to crave sugars. Sugars provide the energy that drives the body's muscles and the brain. In fact, all cells in our body need some sugar to get them going and keep them going. But what do we mean when we talk about 'sugar'?

Observations about sugar include that sugars taste sweet. And things that taste sweet are pleasurable to eat. Babies and children especially crave sweet foods. Breast milk has some sugars in it, so even the first food a newborn may eat provides an instant pleasure beyond simply satisfying hunger. Most of us come into contact with sugar through the solid, white, granular stuff we put on cereal or in coffee. But for most of human civilization, sweetness was derived through fruit juice or honey. The product we call *sugar* is a relatively recent development, at least in the west.

The earliest evidence of human civilization has been dated to around 3500 BCE,[1] in regions of today's Pakistan and India, centered around the Indus River. At this time, humanity began to abandon their nomadic lifestyle and settle down as a community of people with shared responsibilities and divisions of labor,[1] which support the collective society. Within 1000 years, large communities of interconnected peoples were living and evolving societal standards for behavior, economics, and government.

One of the largest and most significant developments of this time was the unification of two kingdoms into one: Egypt was created under King Menes in 3100 BC.[1] During this period, there were many improvements in the general quality of life of the people in the region, despite the authoritarian governmental structure. One area that benefitted most people was the improvements in access to food. Sweet tastes in food were available mainly through the use of honey, grape fruit syrup, and fruits like raisins or dates.[2]

Figure 1 Indus Valley civilization map.

Tupungato/Shutterstock.com

---

[1] http://www.history.com/topics/ancient-history/ancient-egypt accessed March 23,2015.
[2] http://www.reshafim.org.il/ad/egypt/timelines/topics/food.htm accessed March 23, 2015.

But even before this time, indigenous people of Southeast Asia, Indonesia, and New Guinea were growing tall, grass-like plants which, when ripe, could be cut down and chopped into pieces that produced a burst of sweetness when chewed. This sweetness was of a different sort than that enjoyed by Egyptians on the mainland. It was a sweetness that could cure every ailment, according to New Guinean legend, and even served as a kind of mother to the birth of the human race.[3] By 1000 BCE, this sweet sugarcane plant reached the Asian mainland and began its transformation into the solid material called *sugar* today. Arab rulers in Persia (modern day Iran) were especially fond of the sweet substance, now being processed out of the plant and into a sweet powder. As Arab influence spread eastward starting in 500 AD, sugar went along.

Everywhere else in the world was essentially getting sweet flavors the same way the Egyptians had been. Honey was the most common sweetener, but, frankly, most people just lived without adding sweetening substances to food. If one was able to get some honey, it was precious and was used sparingly. Honey is difficult to harvest (bees!), and while being sweet to the taste, it also has the disadvantages of being a sticky liquid that is a challenge to transport and to use in cooking. When solid sugar arrived on the market, it was an obvious improvement to honey: sugar is solid, and as such is able to be transported, stored, and used much more easily than honey. While it is true that honey is actually sweeter than sugar,[4, 5] the ease of using granulated sugar from sugarcane often means that more sugar can be consumed in any one sitting than honey could be, given products could be made with either product.

Interaction between Arab cultures from the east and European cultures from the west started in earnest during the Great Crusades (1095–1291 AD). The Great Crusades were launched as a response to Islamic expansion from the Arabian Peninsula starting in the 7th century AD and reaching the edges of European space by the 11th century.

The perception at the time was that Arab conquerors were negatively impacting European Christian properties and economic interests in the Middle East. As European soldiers representing the Christian forces of the Catholic Church interacted with Arab soldiers and cultures, they inevitably interacted with foods and medicinal substances of those regions and brought back information about what they had seen, used, and tasted to their friends and families in Europe.

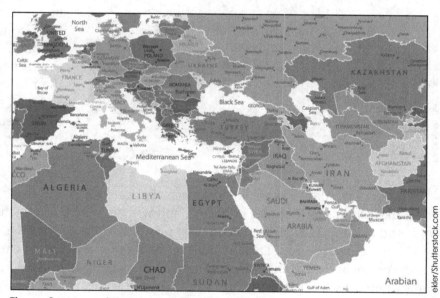

Figure 2   Map of the spread of Islam, 622–750 CE.

---

[3] Cohen, Rich. "Sugar Love (A Not So Sweet Story" *National Geographic*, Volume 224, No. 2, pp. 78–97 August 2013.

[4] http://www.diffen.com/difference/Honey_vs_Sugar accessed March 23, 2015.

[5] Tro, Nivaldo (2009). Chemistry in Focus 4th Ed., Brooks-Cole: Belmont, CA. p. 514.

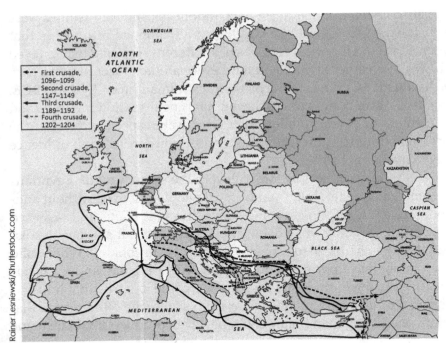

Figure 3   Map of the Great Crusades.

Figure 4

European appetite for sugar was ravenous. The end of the Crusades was not an end to the desire for sugar. Trade routes developed to bring Arabian and Indian sugar to Venice, which was the main port of exchange for European money and exotic goods. Venetian traders exacted exorbitant prices from their sugar patrons, claiming the high costs of logistics for the tropical product. Europeans had already become accustomed to adding sugar to medicinal substances, as well as for more standard cooking and eating uses.[6] They weren't about to give it up now. But the prices were out of reach of everyone except the aristocracy. And as demand for sugar grew past that for honey during the 15th century, along with the rise in price for this labor-intensive product, there came tiptoeing along a resentment of fat Venetian traders and their large profit margins for the coveted sugar. A response was about to be given to the establishment in Venice. A period in history now called the *age of discovery* was beginning, fueled in part by addiction to sugar and a lust for profit.

The tropical plant, which is the source of granulated table sugar, is called *sugarcane*. It is a tall, grass-like plant, up to 12 feet in height, and with stems up to 2 inches in diameter. It is sensitive to cold, and as such, it is more successfully planted in places where the climate does not include temperatures below freezing. This includes parts of the southeastern United States, Mexico, Central and South America, Pacific Islands, and many tropical countries in Africa and Asia. It is the sap of the plant where the sugar is found and from which the sugar must be extracted. Modern harvesting and refining processes make the period between cutting and extraction as short as possible, as a delay after harvesting will result in a drop in sugar content in the plant.[7]

But until the age of mechanization, sugar harvesting had to be done by hand. This is long, hot, hard work. And in order to make a profit, labor costs need to be as low as possible. During the Age

---

[6] LeCouteur, Penny, and Jay Burreson (2005). Napoleon's Buttons. Jeremy M. Tarcher/ Penguin Group: NewYork, NY. p.52

[7] C. Hort (n.d.) "Sugar Cane". Accessed at http://www.hort.purdue.edu/newcrop/crops/Sugar_cane.html on March 26, 2015.

of Discovery, this often meant enslaving indigenous people to grow the cane for Europe's markets or bringing people by force from other parts of the world to work in the sugarcane fields as *Slaves*. The European support of the slave trade was actually waning by the time the Age of Discovery began, but demand for products like sugar, and for profits from products like sugar, reinvigorated the slave trade with a vengeance. European germs often decimated New World native peoples with diseases like smallpox and yellow fever, but African peoples fared better in surviving diseases and were generally more accustomed to the hot climates of the regions where sugarcane was being planted by colonial powers.[8] Therefore, in a short time, a terrible trade route developed, now known as the Great Circuit.

The Great Circuit connected Europe, Africa, and colonies in the Americas. There were essentially three legs to this circuit. The first leg was a trip from European ports to West African ports with finished goods. These goods were used to pay European slave traders and to 'barter' for the exchange of people to go to the New World. History reveals that most of the time, people were simply kidnapped and no goods ever exchanged hands. The 'middle passage' was the most horrific: after marching up to 1,000 miles to ports, slaves were herded like cattle down into cramped holds on ships and transported across the Atlantic to the New World colonies and territories in voyages that lasted two to four months. Several people died in this middle passage from exhaustion, disease, insanity, starvation, or suicide. Some sources say between one and two million African people died during the Middle Passage.[9] After arriving in the New World, captains sold the remaining living slaves and bought materials for the return journey to Europe. This was the last leg of the terrible cycle. Often the slave ships would also carry raw materials like raw sugarcane syrup or cane-based products like rum back in their holds to sell to European manufacturers and clients. Slave traders were wealthy, getting paid for every leg of their journey. The incentive to continue the practice was strong, especially as sugar manufacturers reaped huge profits from slave labor forces subsidized the work of the slave traders. For 300 years, the Great Circuit brought generations of Africans, some estimates suggest more than 50 million people, and some others, to Europe and the Americas against their will. In Europe, after nearly 20 years of campaigning by abolitionist politician William Wilberforce, slavery was finally outlawed in 1803, and all slaves were freed by legal decree in 1833. In the United States, which had been born during the 300 years of Great Circuit operation, slavery was a contentious issue, having been prohibited in founding documents of the young nation, but so many exceptions were made to skirt the antislavery laws that many states were still allowing slavery by the middle of the 18th century.[10] Certainly, history reveals that Southern states, in which sugarcane was a profitable crop, benefitted from the low labor costs that slavery afforded. The outbreak of the Civil War in America (1861–1865) ended slavery with the Emancipation Proclamation, signed by President Abraham Lincoln into law in 1862.

What is it about sugar that would cause all this? The sugarcane plant itself is not unique, in that it is like all green plants: it makes sugar through the process of photosynthesis. Photosynthesis takes the energy from sunlight and uses it to rearrange molecules of carbon dioxide and water into sugar and oxygen:

$$6CO_2(g) + 6H_2O(liq.) \xrightarrow{\text{\textit{Light Energy}}} C_6H_{12}O_6(aq) + 6O_2(g)$$

---

[8] "Europe Before Transatlantic Slavery" (2011). Accessed at http://understandingslavery.com/index.php?-option=com_content&view=article&id=315&Itemid=150 on March 30, 2015.

[9] Africans in America: The Terrible Transformation, Part 1: The African Slave Trade and the Middle Passage (1998) Narrative retrieved from http://www.pbs.org/wgbh/aia/part1/1narr4.html on March 30, 2015.

[10] "Slavery in the United States" (2014). Accessed from http://www.sonofthesouth.net/slavery/slavery-us-constitution.htm on March 30, 2015.

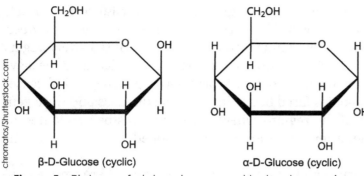

β-D-Glucose (cyclic)   α-D-Glucose (cyclic)

Figure 5 Pictures of alpha-glucose and beta-glucose, in cyclical form.

This is a conversion of solar energy to chemical energy. The energy from the sun is stored in the new bonds created when the sugar, $C_6H_{12}O_6$, is made. This sugar is called a *simple sugar* and it is glucose. Simple sugars are single carbohydrate molecules that can taste sweet. Carbohydrates are molecules of carbon, hydrogen, and oxygen that have a general composition of $C_x(H_2O)_y$ (where x could be different from y).

When glucose is made, it can have one of two different structures, alpha-glucose and beta-glucose. Look at them side by side:

You can see the difference is at the carbon on the right side, the carbon marked '1'. If the −OH group on this carbon is put on below the carbon, it is called *alpha*-glucose, or α-glucose. If the −OH group on carbon #1 is put on above the carbon, it is called *beta*-glucose, or β-glucose. Each type of glucose has special purposes in plant and in human use as well. We'll cover this more in the unit 'Polymer'.

Fructose is another simple sugar produced by plants during photosynthesis, and it can also come in alpha and beta configurations, as shown in Figure 6.

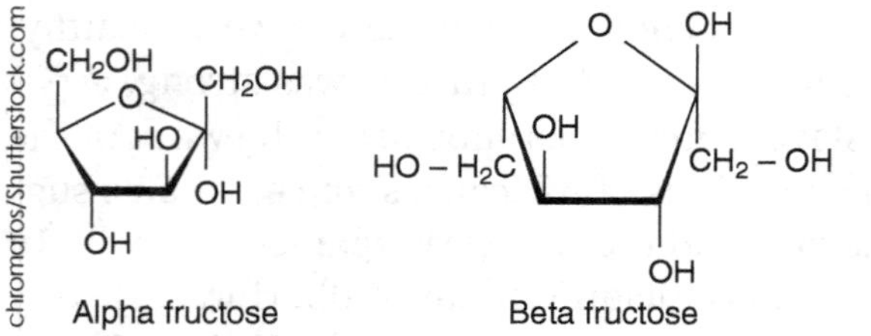

Alpha fructose   Beta fructose

Figure 6 Alpha- and beta-fructose molecules, cyclical configurations.

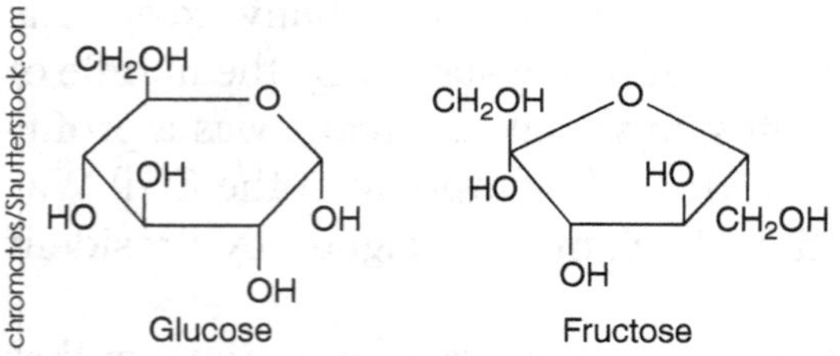

The biggest obvious difference between glucose and fructose is the size of the ring that is formed. Glucose has six atoms in its ring (five carbons and an oxygen atom), and fructose has five members in its ring (four carbons and an oxygen atom).

Right away, we can see that the structures of these sugars are different, but the chemical formulas for both are $C_6H_{12}O_6$. This means that glucose and fructose are *structural isomers*. They are composed of the same elements in the same general ratios, but they are put together differently. Any plant has its own recipe book for how much glucose and fructose to make, and yet, it is the same process that builds each: photosynthesis.

Glucose   Fructose

Figure 7 Comparison of alpha-glucose and beta-fructose molecules.

But sugarcane is mostly composed of another sugar: *sucrose*. Sucrose is the white, granulated material that made Europeans ditch honey all those centuries ago and change the face of the earth with slavery. How is sucrose made in the sugarcane plant? Sucrose is made through a chemical reaction that combines an alpha-glucose molecule with a beta-fructose molecule. The chemical reaction that occurs is called a *condensation reaction*. A condensation reaction, also known as a dehydration synthesis reaction, brings together two smaller molecules to create a larger one, usually by making water as a side product. This is what happens inside the sugarcane plant to create sucrose:

Notice that we can represent this reaction symbolically with words:

Alpha-glucose chemically combined with beta-fructose yields sucrose and water.

Or with symbols and condensed chemical structures:

$$C_6H_{12}O_6 + C_6H_{12}O_6 \rightarrow C_{12}H_{22}O_{11} + H_2O$$

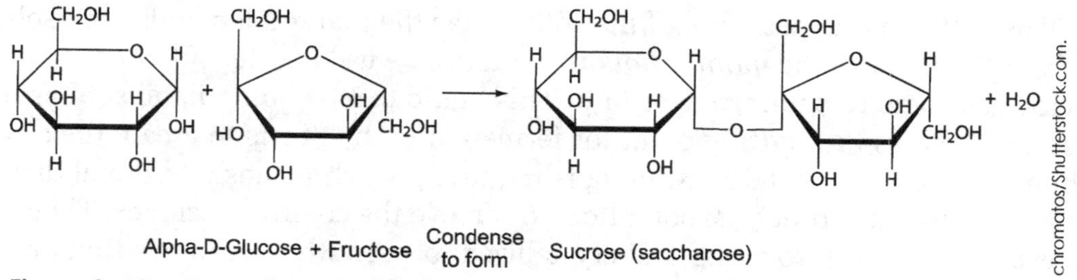

Alpha-D-Glucose + Fructose — Condense to form → Sucrose (saccharose)

Figure 8 Condensation reaction.

To further condense the chemical reaction equation, it could be written as

$$2C_6H_{12}O_6 \rightarrow C_{12}H_{22}O_{11} + H_2O$$

Each representation has its advantages and disadvantages, but each is also equally balanced and correct. The choice of equation format depends on the situation in which it will be used, and the audience which will be receiving it.

Tropical plants like sugarcane benefit from the production of sucrose from glucose and fructose. Sucrose plays a primary role in plant growth and development, as a source of energy and of building materials for structure and storage needs.[11] Release of a water molecule in the production of sucrose is doubly advantageous as sucrose is highly soluble (meaning that sucrose *dissolves well*) in water, and this makes the larger molecule easier to move around the plant. This easy aqueous solubility (ability to dissolve in water) also makes it easy for humans and other animals to use the sucrose that the plant makes when we eat the plant, or to extract the sucrose in order to produce the solid granules of sugar we have come to know and love. Sugarcane production and refining have not changed in essence for the last several hundred years, although the processes by which we accomplish each stage of production and refining has become almost entirely mechanized. As mentioned earlier, most sugarcane production prior to the period known as the Industrial Revolution (ca. 1760–1840 AD in Europe, later in the United States) was done by hand, and in societies where demand was sufficient, slave labor was used to increase production and revenue. Sugarcane is grown by planting sections of mature stalk that sprout and grow new roots and shoots. Once the plant has grown to maturity, it can be cut down and allowed to regrow from the same root two to three more times to produce up to four successive harvests from the same root base.[12]

After harvesting the sugar cane stalks from the field, the canes are brought to a building where the canes are crushed by passing them through heavy rollers that brings the juice gushing out. The leftover plant material, called *bagasse*, makes great fuel for factory furnaces. The extracted juice is treated with calcium oxide (called *lime*) that helps separate soil and other particulate matter from the juice as well as create an environment in the sugar solution, which will prevent the sucrose from breaking down into fructose and glucose, a process called *inversion*.[12,13] At this point, the juice is *reduced* by evaporating off some of the water in the juice with a boiling water bath process. The resulting liquid is thicker and more syrup-like. The final stage involves outright boiling of the remaining syrup until a *supersaturated solution* is created. This is a situation where there is more sugar dissolved in the water while the solution is hot than the water itself would be able to hold when cool. As the solution cools, the sucrose molecules actually crystallize and solidify and separate from

---

[11] Levine, M. Topics in Dental Biochemistry (2011). Springer: Berlin Heidelberg. pp. 17–27. Accessed online March 30, 2015.

[12] "Agriculture and Manufacturing" (2011). United States Sugar Corporation. Accessed at http://www.ussugar.com/sugar/agriculture.html on March 30, 2015.

[13] "How Sugar is Made" (n.d.) Accessed at http://www.sucrose.com/lcane.html on March 30, 2015.

the liquid. This is the first appearance of the solid sugar the market demands. The solid sucrose is separated from the liquid – the *mother liquor* – and can be used at this point as brown sugar. The mother liquor has some remaining sugar in it. This liquid can be bottled and sold as *molasses* for cooking uses, or used as the starting point for fermentation by yeasts to create the alcoholic drink *rum*. If white sugar is desired, more refining is required, which means additional dissolving steps and filtering the solution through carbon filters to remove the coloring residues. Then, the clarified solution is evaporated down to syrup and crystallized to pure white granules that are bagged and sent to market.

Early sugar was brown. But people didn't care about the color. Solid sugar was *delicious*. On the tongue, from the moment the sugar touched it, the message of *SWEET* was registered in the brain and the series of cascading chemical reactions in the brain set the stage for desire for this taste sensation. Our brains and bodies are hardwired for sugar. On the tongue, there are structures called *taste* buds that are dedicated to different taste sensations. Current research indicates that the taste we understand as 'sweet' is just one of five main taste sensations, which also includes salty, sour, bitter, and *umami*, which is a meaty, savory taste.

Taste buds on the front tip of the tongue have a concentration of proteins, which interact with sweet molecules in a way which results in electrical impulses being sent to the brain. One theory of this interaction is called the *A–H/B* model, proposed in 1967.[14] This theory indicates that the special shape of the protein on the surface of the taste bud enables a strong intermolecular attraction to occur with sweet molecules like sucrose. The A–H/B model suggests that structures of –OH or –NH (the 'A-H' part of the model) on the protein are able to interact with oxygen or nitrogen atoms (the 'B' part of the model) in sweet molecules; likewise, the sweet molecules have –OH or –NH groups, which interact with oxygen or nitrogen atoms on the taste bud proteins.

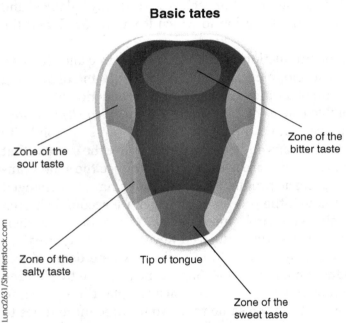

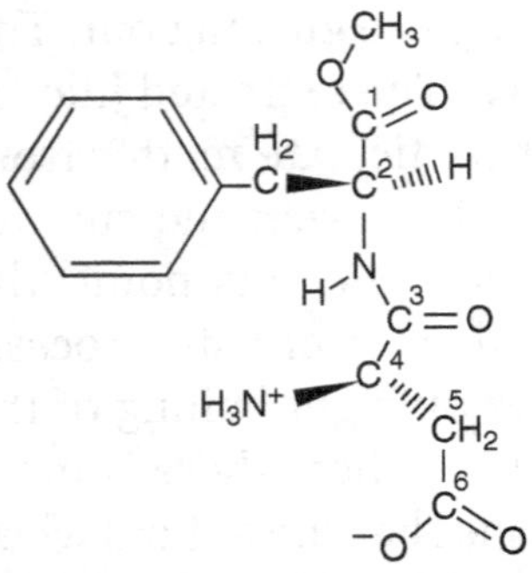

Basic tates

Zone of the sour taste

Zone of the bitter taste

Zone of the salty taste

Tip of tongue

Zone of the sweet taste

Figure 9 Picture of tongue with taste regions, inset with taste bud structure.

Figure 10 Attachment diagram for A–H/B theory of three-point attachment.

This mutual attraction completes a type of electrical circuit and is possible through an interaction known as *hydrogen bonding*. Hydrogen bonding is classically defined as the attraction of a hydrogen bonded to an oxygen, nitrogen, or fluorine atom for an adjacent oxygen, nitrogen, or fluorine atom. This attraction is not a true bond, as understood to be a sharing of electrons as we see between carbons in an organic molecule, for example; but is a strong electrostatic force drawing the electron-poor hydrogen atoms on one molecule toward

[14] DuBois, Grant E. "Unraveling the Biochemistry of Sweet and Umami Tastes." Proceedings of the National Academy of Sciences of the United States of America 101.39 (2004): 13972–13973. PMC. Web. 30 Mar. 2015.

the electron-rich oxygen, nitrogen, or fluorine atoms nearby. This attractive force is enough to hold molecules very tightly together. For example, the surface tension of water that supports insects on the surface of a pond is caused by hydrogen bonding between water molecules.

Hydrogen bonding holds the strands of DNA together.

The strength of ice to support weight in winter and yet be light enough to float is due to hydrogen bonding between water molecules. Some authors suggest that the evolution of life would not have happened if ice didn't float. "Adenine and thymine are connected with two *hydrogen bonds*" and "Cytosine and Guanine are connected with *three hydrogen bonds*".

Figure 11 Picture of a water strider on the surface of a pond.

So, one might conclude that without hydrogen bonding, you might not be reading this today. Hydrogen bonding is a huge force in our water-based world and its living systems. We will see hydrogen bonding again and again in this book.

The structure of DNA double helix

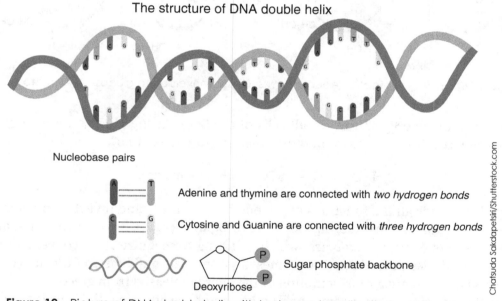

Nucleobase pairs

Adenine and thymine are connected with *two hydrogen bonds*

Cytosine and Guanine are connected with *three hydrogen bonds*

Deoxyribose

Sugar phosphate backbone

Figure 12 Picture of DNA double helix with hydrogen bonding indicated.

Figure 13 Picture of ice on a pond with inset of molecular structure of ice showing hydrogen bonds.

The solubility of sucrose in water is another example of hydrogen bonding at work. The water molecules' hydrogen atoms are attracted to sucrose's many oxygen atoms, and the hydrogen's on sucrose are attracted to water's oxygens as well. This mutual attraction accounts for the stability of sucrose in water, whether it's inside the sugarcane plant, inside a glass of iced tea, or inside of you.

Once on the tongue, the idea that something sweet is there is immediate. Often this produces a pleasurable feeling, caused by the brain's release of neurotransmitter molecules of dopamine,[15] serotonin (regulates mood), and endorphins (reduce pain and help moderate the effects of stress) in the brain. For some, this pleasurable feeling is so intense that it can cause cravings like those associated with other kinds of addictions. There is some evidence to suggest that sugar produces similar stimulation to that of drugs like opium. We will talk about this more in the unit 'Drug'.

Why would humans crave sucrose? Because sucrose is a source of glucose, and glucose is the main fuel for our brains and muscles. When sucrose is consumed, the act of digestion at the beginning of the small intestine includes cleavage of the molecule into alpha-glucose and beta-fructose by reversing the condensation reaction in a process called *hydrolysis*. This action separates the simple sugars from each other, and allows them to leave the small intestine and diffuse into the blood stream for transport through the body.

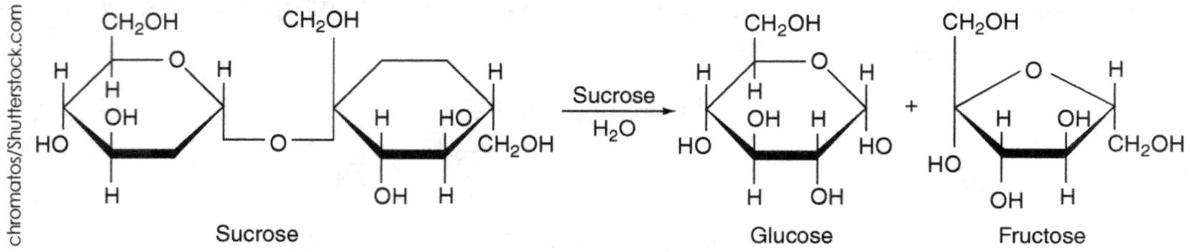

Figure 14 Hydrolysis reaction equation for sucrose conversion to glucose and fructose.

Alpha-glucose is the energy source for all cells of the body. Using glucose for energy is called *cellular respiration*. It is the opposite of photosynthesis; can you see this?

$$C_6H_{12}O_6 + 6O_2 \rightarrow 6CO_2 + 6H_2O + energy$$

This energy is used for all the functions of the body, both visible and invisible: moving muscles and joints, keeping us warm, remembering ideas and recognizing faces, repairing injuries, and much more. We measure this energy in *calories,* which you might recognize from conversations about food. In reality, the calories in food are 1,000 times larger in scale and labeled *Calories* (with a capital C), but in essence, both are a representation of the energy released from glucose.

Which begs the question, what if someone's body cannot use glucose well? This is the condition commonly known as *diabetes*. Diabetes is a common ailment, affecting 29.1 million Americans in 2012.[16] Of this number, 1.25 million have type 1 diabetes, also known as Juvenile Diabetes or *diabetes mellitus*. The remainder of diabetics is diagnosed with type 2 diabetes, also known as adult-onset diabetes. These two diseases have some similarities and also some crucial differences, summarized in the table below. Both are diseases caused by problems with insulin, a molecule made by special cells in the pancreas – *the islets of Langerhans*, also known as beta cells. Insulin is tasked with opening cells to receive glucose. Without insulin, no cell can get glucose or any energy from it. Without insulin, the only option is death.

---

[15] Paradowski, Robert J., PhD. 2012. "Sugar addiction." *Salem Press Encyclopedia Of HealthResearch Starters*, EBSCOhost (accessed March 30, 2015).

[16] "Statistics about Diabetes" (2014) from the National Diabetes Statistics Report, 2014. Accessed from http://www.diabetes.org/diabetes-basics/statistics/ on March 30, 2015.

COMPARISON OF TYPE 1 AND TYPE 2 DIABETES		
Characteristic	Type 1 Diabetes	Type 2 Diabetes
Age at onset	Any age is possible, though most commonly diagnosed in children, adolescents, or young adults	Any age is possible, most common over age 45, often also overweight
Insulin production	Insulin production falls very low then usually ceases completely. Usually, an immune response destroys the body's own pancreatic cells	Insulin is made at normal or near-normal levels
Cellular insulin use	Insulin can be used efficiently by cells if present	Insulin is not used efficiently by cells
Treatments	Delivery of insulin via injection. Research is aimed at 1) finding ways to transplant pancreatic cells that will restart insulin production; 2) finding ways to effectively deliver insulin mechanically through external means	Diet and exercise, reducing weight, adding drugs to either stimulate insulin production or inhibit glucose uptake by cells.
Symptoms at onset	Symptoms can develop over days or weeks. Excessive thirst and urination, rapid weight loss, incessant hunger, blurry vision, blood sugar values usually well above normal limits (blood glucose levels > 120 mg/dL). Acute symptoms include vomiting, stomach pain, a strong, fruity breath odor, flushed but dry skin, and coma. Death can occur	Blood levels above 120 mg/dL, often developing over months or years. Symptoms similar to type 1, though much usually less acute in developing. Secondary symptoms include slow healing of cuts or sores, tingling in hands and feet (diabetic neuropathy).

So glucose is the magic that makes life go. What about fructose? What is its purpose? Fructose is made in equal quantities as glucose in order to make sucrose in sugarcane. What happens to fructose when hydrolysis splits the sucrose molecule in the small intestine? Fructose is used in the respiration process and is also present when creating large molecules called *glycogen* (see the unit 'Polymer'), which store glucose for emergency use. Fructose is also a major player in fat production for long-term energy storage for the body.

Which means that fructose also causes some problems. Excessive consumption of fructose is implicated in insulin resistance (type 2 diabetes), obesity, and liver disease.[17] Development of products like high-fructose corn syrup and the use of this product in many convenience foods has led some to the conclusion that this product has been a major player in America's obesity epidemic.

Sugars like glucose, fructose, and sucrose are common in plants. Some plants make more of one kind of sugar than another, though most make some of all three at least. Vegetables have sucrose, and glucose, some fructose, and polymers of glucose called *starches*. Fruits have sucrose and glucose and usually more fructose than vegetables. Notice that all words end in -ose. You can recognize a sugar when it has this suffix. In a product like high-fructose corn syrup, the starch in corn, which is a polymer containing only alpha-glucose, is broken down into its monomer units and then many of those alpha-glucose molecules are converted to fructose molecules. On the tongue, fructose tastes sweeter than sucrose and glucose, and industrially high-fructose corn syrup is a more efficient sweetener than glucose or sucrose. This fructose stimulates the sweet receptors on the taste buds, creates a dopamine pleasure response in our brains, and reinforces the desire for sweet foods. It has also been shown through research that fructose stimulates the

---

[17] "Fructose vs Glucose" (n.d.) Accessed at http://www.diffen.com/difference/Fructose_vs_Glucose on March 30, 2015.

hunger response.[18] There is high correlation that consumption of foods and drinks high in fructose (containing high-fructose corn syrup) is contributing to the obesity problem in America.

Artificial Sweeteners

For some, the solution to obesity is dieting, and dieting can include removing some things from the diet in order to cut calories. For many dieters, this means replacing sugar with a sugar substitute that has few or no calories to contribute to the diet. But the molecule must still taste sweet, so it must have some structures that can interact with the sweet taste bud receptors and send the sweet signal to the brain for positive identification. We have many artificial sweeteners on the market, though some are older and some are sweetener than others. They are not all the same.

Many people identify the most common artificial sweeteners by the color of the packets, which stand on restaurant tables. The pink paper is associated with the oldest commercial artificial sweetener, called *saccharin*. Developed in 1878 at Johns Hopkins University by researchers investigating sulfobenzoic acids, saccharin's sweet taste was discovered by accident when one researcher brought some of his lab work home on his hands. When his dinner roll tasted extraordinarily sweet, the scientist Constantin Fahlberg ran back to the lab to taste everything on the lab bench (yikes!) until he found the source of the sweet taste. It was a compound called *benzoic sulfinide* – the substance we know as saccharin.[19]

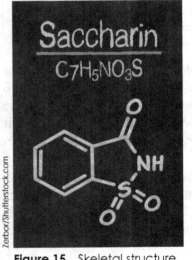

Figure 15   Skeletal structure of saccharin.

Saccharin is much sweeter than sugar. It is considered 300 times sweeter than sucrose. Within 10 years of its discovery, saccharin was being produced and sold to a willing public. Manufacturers of processed food saved money by using saccharin instead of sugar to sweeten their foods. Dieters were thrilled to have a way to cut calories without having to give up sweet flavors. At the same time, concerns were growing about the negative health effects that saccharin might have on users. As the first half of the 20th century unfolded, saccharin enjoyed ups and downs in public demand and in governmental regulation. But as the world emerged from the Second World War, things changed. Saccharin use usually decreased after wartime sugar rationing was lifted. But not after World War II. The new availability of home refrigeration and more processed food products kept saccharin producers busy and consumers happy. However, changes in scientific inquiry reopened questions about saccharin's potential negative health effects. By the 1970s, a few more artificial sweeteners were on the market, and saccharin was off. Longer term animal studies suggested a link between saccharin use and increased risks of bladder cancer. But after two more decades of research, the link has not been proven, so saccharin is still available for use today.

What other artificial sweeteners are on the market? Saccharin's successes inspired other scientists. Aspartame, often marketed under the name Nutrasweet, was created in 1965 and on the

---

[18] Lane, M. Daniel and Seung Hun Cha. (2009). "Effect of glucose and fructose on food intake via molnyl-CoA signaling in the brain." *Biochemical and Biophysical Research Communications*, Vol. 382, Issue 1, 24 April 2009. Pp. 1–5. Accessed on line at http://www.sciencedirect.com/science/article/pii/S0006291X09004306 on March 30, 2015.

[19] Hicks, Jesse. "The Pursuit of Sweet: The History of Saccharin". *Chemical Heritage Magazine*, Spring 2010. Accessed online at http://www.chemheritage.org/discover/media/magazine/articles/28-1-the-pursuit-of-sweet.aspx on March 31, 2015.

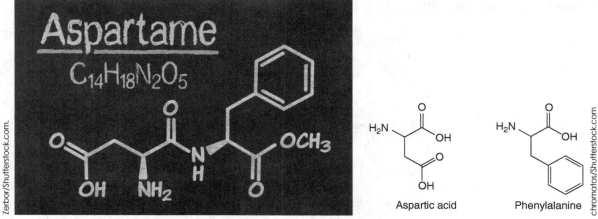

Figure 16 Skeletal structure of aspartame.

commercial market in 1996. Aspartame is about 200 times sweeter than sugar and has the ability to be metabolized (broken down in the body) into aspartic acid, phenylalanine, and methyl esters or methanol. This is an advantage for some because aspartic acid and phenylalanine are amino acids that are generally used by the body anyway and therefore toxic side effects are minimal.

But all is not calm in Camelot. Aspartame has a dark side, and it is connected with a genetic condition called *phenylketonuria* (PKU). This disease affects one in every 10,000–15,000 newborns in the United States each year and is characterized by an inability to breakdown phenylalanine into usable products in the body. As a result, people with PKU must avoid products that contain phenylalanine, as phenylalanine is a toxin for patients with PKU and can cause significant brain damage if left undetected. In addition, it is also true that methanol is a poison in high doses in the body because of conversion of methanol to formic acid,[20] but the small amounts created by the metabolism of aspartame in normal doses are generally not harmful.

There are also other artificial sweeteners. Sucralose, initially marketed under the commercial name *Splenda*, is a more recent player. Developed in 1976 and approved in 1998 by the Food and Drug Administration (FDA) for general consumption, aspartame is a sweetener whose structure should look very familiar to you.

Figure 17 Skeletal structure of sucralose.

Sucralose (E955)

Sucrose (saccharose)

Early ad campaigns for Splenda included a tag line: 'Made from sugar, so it tastes like sugar'. This was a nod to the fact that the creation of sucralose does, in fact, begin with sucrose. But that is really where the similarity ends. There is a lot of manufacturing that goes on to exchange three –OH groups on sucrose for three chlorine (Cl) atoms. This exchange changes *everything*. Sucralose is

[20] Liesivuori, J. and Savolainen, A. H. (1991), Methanol and Formic Acid Toxicity: Biochemical Mechanisms. Pharmacology & Toxicology, 69: 157–163. doi: 10.1111/j.1600-0773.1991.tb01290.x.

600 times sweeter than sucrose. Sucralose is not metabolized in the body, so there are no calories in it. Dieters can smile with sucralose among the options.

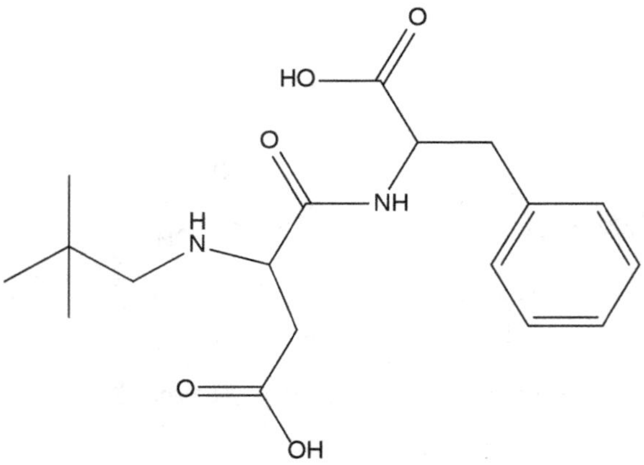

Finally, there is now *neotame*, which is 8,000 times sweeter than sugar. Neotame is the youngest member of this group of artificial sweeteners. Created in and approved for marketing in 2002, neotame is similar to aspartame in that the amino acids aspartic acid and phenylalanine are involved, and methanol, but in neotame, the compound is a 'derivative'[21] of the dimer composed of these two amino acids. The structure of neotame is shown below:

Figure 18   Skeletal structure of neotame.

The structure of neotame, though composed of the same basic components of aspartame, does not appear to break down into it's constituent parts during metabolic processes, and thus is apparently not a risk for phenylketonurics. In fact, the amount of neotame used in most products is so small (because the sweetener is so powerful) that a warning to phenylketonurics is not required. The use of neotame in commercial foods is growing, and for some, this represents a real breakthrough because this artificial sweetener possesses all the characteristics desired: incredible sweetening power, minimal toxicity, and wide application. Time will tell how neotame performs against other artificial sweeteners.

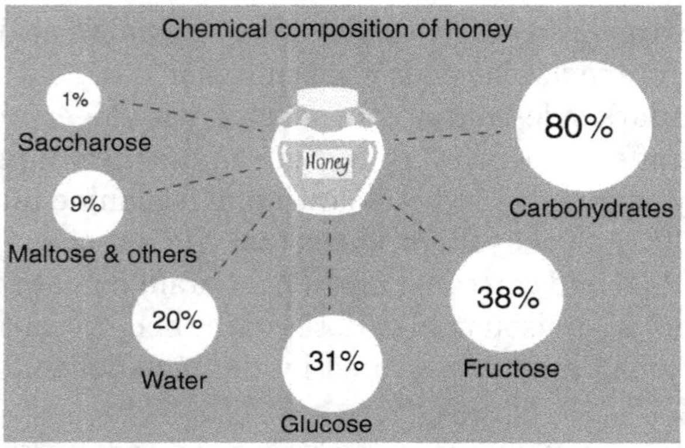

Figure 19   Pie graph of composition of honey.

Artificial sweeteners are just one classification for sweetener compounds. There are also natural sweeteners, dietary supplements, and sugar alcohols.[22] Natural sweeteners are defined as being those substances that are naturally sourced, like sucrose is obtained from the sugarcane plant. Other examples of natural sweeteners include honey, maple syrup, agave nectar, molasses, corn syrup, and so on. These sweeteners contribute calories but are generally less processed than artificial sweeteners. These natural sweeteners are more than just one compound; they are mixtures of compounds in water. Just as an example, look at honey's chemical composition:

It is clear that more than one substance makes up honey. *Mixtures* are combinations of compounds that are not chemically joined to each other, but exist together in the same larger substance. In the case of honey, it is classified as a *homogenous mixture*, because homogenous means that all the substances are evenly distributed throughout any given sample of honey. Visually speaking, homogenous mixtures look uniform, and a small sample looks identical to a large sample.

The alternative to a homogenous mixture is a *heterogenous mixture*, and this is different from a homogenous mixture because members of the mixture can be distinguished out of the mixture from

---

[21] "Neotame Fact Sheet" (2008). Accessed via http://neotame.com/about.asp on March 31, 2015.
[22] Anderson, Pam. "How Sugar Substitutes Stack Up". *National Geographic*, published July 17, 2013. Accessed at http://news.nationalgeographic.com/news/2013/07/130717-sugar-substitutes-nutrasweet-splenda-stevia-baking/ on March 31, 2015.

other members. A common example would be fruit salad. Fruit salad is often composed of pieces of melon, whole grapes, slices of citrus fruit, and maybe even pitted cherries. The composition of the fruit salad is visible because the pieces of the mixture are large enough to identify by sight.

One feature of all mixtures is that because they are physically combined (not chemically combined), the members of the mixture can be separated by physical means from one another. In some cases, this is easier than others; for example, separating the types of fruit from one another in a fruit salad would be easier than separating the components of a sample of honey, but it is possible. Physical properties like texture, color, boiling point (the temperature where a substance passes from being a liquid to being a gas), melting point (the temperature where a substances passes from being a solid to being a liquid), freezing point (the temperature where a substance passes from being a liquid to being a solid), solubility (the ability or ease of a substance to dissolve in another substance), and

Figure 20 Picture of a bottle of honey, with an inset of an up-close picture of honey, looking the same as the larger sample.

Figure 21

density (the amount of mass present in a given volume of that substance) are all unique to each pure substance in a mixture, and therefore, it can be used to separate those substances from each other as needed.

Dietary supplements, on the other hand, are substances that are usually also often naturally sourced but are still processed to become solid sweetener products. The FDA defines dietary supplements as those substances that are eaten and that provide something supplemental (the implication is *beneficial*) to the diet.[23] This beneficial supplement can include 'vitamins, minerals, herbs or other botanicals, amino acids, and substances such as enzymes, organ tissues, glandulars, and metabolites'.[24] Supplements do not have a defined physical form or delivery types and are not regulated specifically by the FDA; instead, makers of dietary supplements are required to know and adhere to the FDA requirements for reporting and packaging, with the understanding that if the maker is caught being fraudulent, they are subject to prosecution by the FDA to the fullest extent of the law. One thing that is always true about dietary supplements, though, is that they are considered foods, *not drugs*. Examples of dietary supplements, which are sweeteners, include Stevia and monkfruit extracts.

Stevia is a product of the leaves of the tropical rainforest plant *Stevia rebaudiana* plant, originally

---

[23] "Q & A on Dietary Supplements" (2014). Accessed at http://www.fda.gov/Food/DietarySupplements/QADietarySupplements/default.htm on April 6, 2015.

found in the rainforests of Paraguay in South America. Chewing the leaves for their sweet taste, or using the leaves in teas or local medicines as a sweetener, was common among indigenous peoples of the eastern mountain region when Spanish explorers entered the area in 1537.[24] By the time South American colonialization was nearly complete in the 19th century, using the herb was common in many countries in South America.[25] Sugar producers in America were aware of the sweet product by the end of the 18th century, but they were threatened by the potential negative impact this exotic sweetener might have on the market share enjoyed by sucrose manufacturers, so its introduction to the general public as a sweetening product would take another 100 years. The plant creates several sweet compounds, but the most prevalent one is *steviol*. This is the most common sweetening compound sold as the Stevia product.

Leonid andronov/Shutterstock.com.
Swapan photography/Shutterstock.com

Figure 22   Picture of a *Stevia rebaudiana* plant, with an inset of the steviol compound.

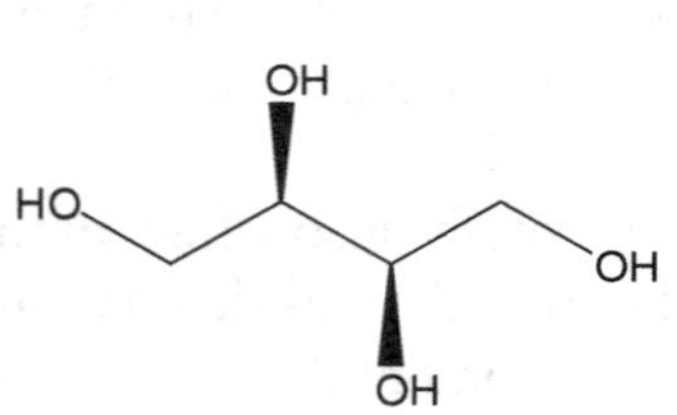

molekuul.be/Shutterstock.com
Courtesy of Author

Figure 23   Skeletal drawings of sugar alcohols erythritol and xylitol.

The final group of alternatives to natural sweeteners is *sugar alcohols*. This group of molecules is rather large, often being created from sugar polymers like cornstarch through the action of microorganisms like yeast, which break down the larger sugars through fermentation processes. Generally speaking, sugar alcohols are chains of carbons with many –OH groups of the chain. This includes erythritol, xylitol, mannitol, glycerol (also known as glycerin), isomalt, lactitol, maltitol, and sorbitol.[26] These molecules have sweet tastes, though not as sweet as classic sugar (sucrose), and, when metabolized in the body, also produce calories. So they are not the best option for everyone. Two of these sweeteners, xylitol and erythritol, are especially popular with dentists because these sweeteners will not promote tooth decay the way sucrose, glucose, and fructose do.

[24] "History [of Paraguay]" (n.d.) Accessed at http://countrystudies.us/paraguay/2.htm on April 6, 2015. Source: U.S. Library of Congress.

[25] Gates, Donna (2000) The Stevia Story: A tale of incredible Sweetness and Intrigue. Obtained on April 6, 2015 in part from "History of Use" (n.d.), available at http://www.stevia.net/history.htm

[26] "Sugar Alcohols" (2014). Accessed at http://www.diabetes.org/food-and-fitness/food/what-can-i-eat/understanding-carbohydrates/sugar-alcohols.html on April 6, 2015.

Tooth decay and cavities (known as *caries* by dentists) are caused by the action of lactic acid produced as a side product by bacteria in the mouth, which digest sugar. Specifically speaking, the corrosion of tooth enamel (mainly the calcium-containing mineral compound *hydroxyapatite*, Ca_{10} $(PO_4)_6(OH)_2)$[27]) is caused by lactic acid, which is the product of anaerobic (occurs without the aid of atmospheric oxygen) fermentation of simple sugars like glucose and fructose most often by a species of *Streptococcus* bacteria known as *Streptococcus mutans*.[28] The first step in the mouth is the action of enzymes to hydrolyze (add water to split the molecule) sucrose into glucose and fructose, and then the bacteria get ahold of some of these simple sugar molecules and break them down in order to rapidly release some of the energy in those sugar bonds, which bacteria then use for their own life processes.

$$C_{12}H_{22}O_{11} + H_2O \xrightarrow{\text{\textit{sucrase enzyme action}}} 2C_6H_{12}O_6$$

$$C_6H_{12}O_6 \rightarrow CH_3COCO_2H + 2H_2O + \textit{energy}$$

Pyruvic acid

Pyruvic acid → lactic acid

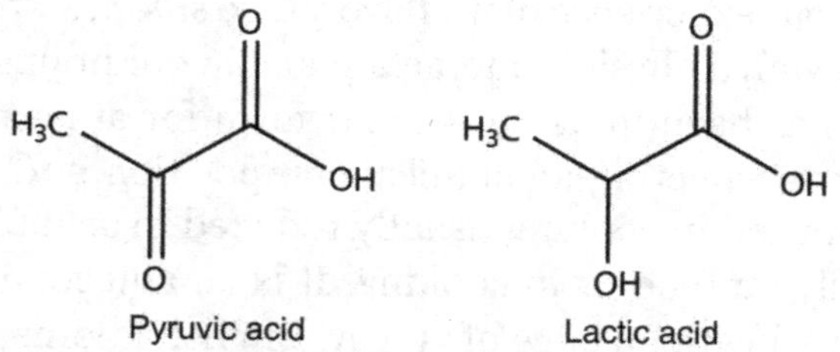

Figure 24 Skeletal structures of pyruvic acid and lactic acid.

chromatos/Shutterstock.com

[27] Loesche WJ. Microbiology of Dental Decay and Periodontal Disease. In: Baron S, editor. Medical Microbiology. 4th edition. Galveston (TX): University of Texas Medical Branch at Galveston; 1996. Chapter 99. Available from: http://www.ncbi.nlm.nih.gov/books/NBK8259/. Accessed on April 7, 2015.

[28] Bisla, Sharan (2000). "Dental Caries – A Student Project". Available at http://sciweb.hfcc.net/biology/jacobs/micro/caries/caries.htm#3. Accessed on April 6, 2015.

SALT

What is sweet without salty? A conversation about food must necessarily include a discussion about salt. The word *salt*, like sugar, is so commonplace that nobody even thinks about what it means. We all understand the word based on our experience of it's flavor. Like *sweet, salty* is a taste sensation that our biology registers on the tongue with a physical interaction and in our brains with a chemical response.

The word *salt* in English has been in use in similar form for at least 1000 years.[29] It is used as a noun, both as a general term reflecting a chemical composition and as a specific term meaning the white, crystalline material used in cooking usually referred to as table salt. It is a verb, implying the action of using salt, usually on food or in cooking. It is an adjective, *salty,* implying something savory or perhaps sharp tongued with an edge of vulgar. English idioms use salt and imply its power to change a situation for the worse (*rub salt into someone's wounds*); to say that someone has an unforgettable character (e.g., '*he's quite an old* salt') or to imply wisdom (to take an idea *with a grain of salt*) or imply value (*worth one's salt*). The Latin root for salt, *sal* (pronounced 'sale'), shows up in some town names where salt mines were located; for example, Salzberg (German word for salt is *Salz*), and Nowa Sól (Polish for *New Salt*). English words like *saline, salinity, and salary* reference salt's presence or it's monetary value in ancient times. The Greek word for salt, *hals*, shows up with words like *halogen* (describing the nonmetal elements that form binary salts with metals) and *halite* (a geology term describing the crystalline rock form of sodium chloride – table salt).

Ask an average person on the street what salt is and they will most likely say that it is the white stuff used in cooking to add flavor. The most common material used for this purpose is sodium chloride, with the formula $NaCl$. This material is a perfectly good representation for salt, though *salts* can be more than just Na and Cl. To chemists, a salt is any material chemically composed of metallic and nonmetal ions, held together by an electrostatic attraction of opposite charges. This might seem confusing at first, but it really is amazing.

First, we have to understand what it means to be *metallic* or *nonmetallic*. Certainly, the best place for this to start is with the periodic table.

Most of the elements in our universe are classified as metals, and as such are printed on the left side of the stairstep line dividing the periodic table one large part and one small part. *Metals* are those materials that have metallic physical characteristics; that is, they are usually solids at room temperature (exception: mercury, Hg, which is a liquid at room temperature), are lustrous (shiny when polished), are

---

[29] "salt". Dictionary.com. *The American Heritage® Dictionary of Idioms by Christine Ammer.* Houghton Mifflin Company. http://dictionary.reference.com/browse/salt (accessed: April 08, 2015).

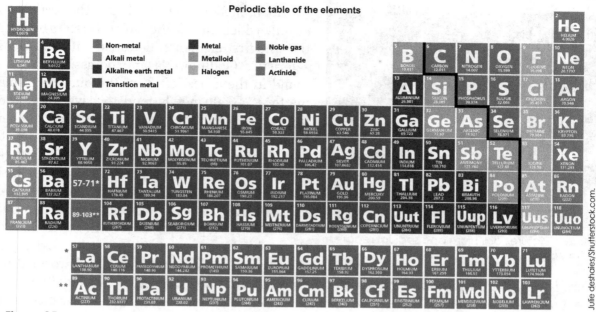

Periodic table of the elements

Figure 25

malleable (able to be shaped without breaking), ductile (can be stretched out into thin wires), conduct (carry) heat and electricity well, have high melting and boiling points, and have high densities.[30] Add to this the chemical properties of metal atoms to give away electrons in chemical reactions and the behavior of metals becomes a little clearer. Metals occupy the entire left side of the periodic table, as well as the two rows beneath the main body of the table. As with any large group, there is physical and chemical variety among the members. Specifically, the metals are grouped into some subclasses:

Region	Member Elements	Common Name
Group 1A	Hydrogen (H)* to Francium (Fr)	Alkali metals
Group 2A	Beryllium (B) to Radium (Ra)	Alkaline earth metals
Groups IIIB to IIB	Scandium (SC) to Copernicum (Cn) – atomic numbers 21–30; 39–48; 72–80; 104–112	Transition metals
	Lanthanum (La) to Lutetium (Lu)	Lanthanide inner transition metals/ rare earth elements
	Actinium (Ac) to Lawrencium (Lr)	Actinide inner transition metals
	Boron (B), Silison (Si), Germanium (Ge), Arsenic (As), Antimony (Sb), Tellurium (Te), and Polonium (Po)	Semimetals or metalloids
Group 3A[31] (not Boron) and others	Aluminum (Al), Gallium (Ga), Indium (In), Thallium (Tl), Tin (Sn), Lead (Pb), Flerovium (Fl), Bismuth (Bi), Livermorium (Lv), and Elements 113 and 115	'Poor' metals[32] or post-transition metals

*Hydrogen is an interesting member of this group. By classic definition, it is not a metallic element, in that it does not exhibit the physical characteristics of a metal, but it is a reactive element in a similar way to the reactivity patterns observed in all the other members of the alkali metals group. This is because the arrangement of the electron(s) involved in that chemical reactivity are very similar among all the members of the group. More on this in a bit. Keep reading.

[30] "Properties of Metals" (2009). http://propertiesofmetals.com/physical-properties-of-metals/ (accessed April 8, 2015).

[31] Boudreaux, Kevin A. (n.d.). "Group 3A" Available at http://www.angelo.edu/faculty/kboudrea/periodic/ periodic_main3.htm. Accessed April 13, 2015.

[32] Jackson, Tom (2012). The Elements: An Illustrated History of the Periodic Table. Shelter Harbor Press, New York, p. 125.

Figure 26 Pictures of lithium and sodium reacting with water.

Alkali metals are named after the Arabic word *al-qali,* which means 'the ash'; this refers to the fact that the first elements identified with these characteristics (sodium, Na and potassium, K) were found in the ashes left from the burning of plant matter.[33] Alkali metals tend to have lower melting points than other metals, they tend to chemically react spontaneously with oxygen and water, and many of the halogen nonmetal elements (see *Nonmetals*) to form compounds that are very stable themselves. The compounds made when alkali metals react with water are slippery to the touch, feeling like soap. In addition, these compounds themselves react spontaneously with acids, which indicates that the metal compounds formed with water are *bases.* Even within the group itself, there is a continuum of reactivity. Reactivity tends to become more spontaneous and more violent for the alkali metal member elements as we move down the group from Lithium to Francium.

Alkaline earth metals are grouped to the right of the alkali metals. Their close proximity to each other indicates a similarity of physical and chemical characteristics, but the alkaline earth metals are slightly less reactive and have slightly greater densities, greater hardness, and higher melting points than the alkali metals. The name *alkaline earth* is a nod to the high prevalence of compounds, which contain these metals in soil deposits. Alkaline earth metals also create compounds that act like bases when reacted with water and burn brightly in oxygen to create compounds of metal oxides. This last property makes magnesium a popular component of fireworks.

Transition metals are located in the rectangular region to the right of the alkaline earth metals. It is essentially 40 elements that have varying metallic properties because of electrical arrangement within the metal atoms themselves. Composed of 4 rows of 10 elements each (Scandium (Sc) to Zinc (Zn), Ytterbium (Y) to Cadmium (Cd), Lanthanum (La) to Mercury (Hg), and Actinium (Ac) to

Figure 27 Picture of elemental magnesium and of magnesium fireworks burning.

---

[33] McQuarry, Donald A; Rock, P.A; Gallogly, E.B. (2011). General Chemistry, 4th Ed., University Science Books, Mill Valley, CA. p. 86.

Ununbium (Uub)[34]), transition elements are exactly what their name implies: this region transitions from more metallic to less metallic characteristics, moving left to right across a row. This variety gives the atoms a great deal of flexibility in bonding. Many transition metals occur in compounds that make excellent pigments for paints. For example, *cobalt violet deep* is the blue–violet compound $Co_3(PO_4)_2$ [cobalt (II) phosphate], and *cadmium orange* is the compound Cd_2SSe [cadmium(II) sulfoselenide].[35]

In addition, transition metals are often critical in molecules that contain organic, nonmetal portions held in specific spatial arrangements by metal atoms at the center of the organic portions. These *organometallic complexes* are often highly colored and have very valuable applications. For example, the substance responsible for the ability to carry oxygen in red blood cells is called *hemoglobin* and is composed of organic sections arranged around an iron atom:

Iron atoms, like many of the atoms of transition metals, have the ability to have several electronic arrangements, and each arrangement suits well for certain purposes. The reason for this is the flexibility of electron arrangement around the nucleus of an atom. More on this in a moment.

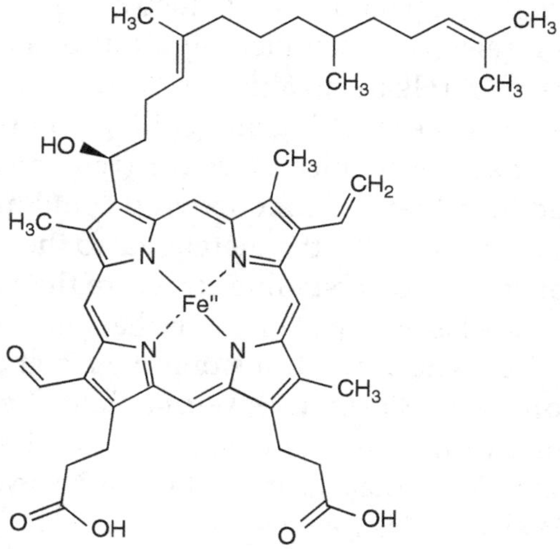

Within the transition metals are two subgroups: the Lanthanide series (Cerium (Ce) to Lutetium (Lu)) and the Actinide Series (Thorium (Th) to Lawrencium (Lr)). These are classified as Inner Transition elements because they are a section of 14 elements (in each series), beginning with the elements that immediately follow Lanthanum and Actinium, respectively. They are located in a separate rectangular section below the main body of the periodic table. The Lanthanide series metals are also known as rare earth metals because of their relatively low natural concentration on earth. Rare earth metals sometimes include Scandium and Yttrium as well because these metals are often found together in mineral deposits.[36] Notice how Scandium, Yttrium, and Lanthanum exist together in a vertical arrangement on the periodic table, which underscores their physical and chemical similarities. Rare earth elements have very valuable uses in: rechargeable cell phone batteries, and hybrid and electric vehicle batteries; as polishing compounds; in night vision goggles, precision-guidance systems for weapons, and as amplifiers for fiber-optic communications systems, just to name a few.[34] They are particularly interesting, and many are physically unstable, meaning that the atoms will decompose into smaller particles over time.

Figure 28 Hemoglobin molecular structure.

---

[34] Ununbium is part of a group of man-made elements which fit the structural expectation that the periodic table's patterns dictate. The name Ununbium is Latin for 112, which is the atomic number for this atom. Ununbium and other man-made atoms have low atomic stability; for example, ununbium will exist in it's elemental form for only 280 ms before decomposing (ref: http://www.chemicalelements.com/elements/uub.html). The action of creating new elements which the periodic table's patterns predict is more an exercise in theory than in practicality. The use for elements with such limited stability has yet to be proven, in many cases.

[35] "Inorganic Pigment Compounds-The Chemistry of Paint" (March 21, 2014). Available at http://www.compoundchem.com/2014/03/21/inorganic-pigment-compounds-the-chemistry-of-paint/. Accessed April 13, 2015.

[36] King, Hobart (n.d.), contributor to " REEs - Rare Earth Elements and their Uses" Available at http://geology.com/articles/rare-earth-elements/. Accessed April 13, 2015.

Like the elements in the alkali and alkaline earth groups, the elements that are arranged vertically with one another in the transition and inner transition elemental regions have similar physical and chemical properties, and the properties become less metallic as the elements progress from left to right across the horizontal rows.

Returning to more metallic behavior is the 'poor metals', otherwise also known as the post-transition metals. This collection includes group 3A metals, which includes Aluminum (Al), Gallium (Ga), Indium (In), Thallium (Tl), and Element 113, currently identified as Ununtrium (Uut). Atoms of these elements tend to behave more like alkali and alkaline earth metals in the way that they behave electronically to form compounds. Similarly, the metals of group 4A include Tin (Sn), Lead (Pb), and Element 114 – Flerovium (Fl) – are similar in their electronic tendencies, though they also have more than one stable electronic state, which makes them a little like transition metals also. Group 5A has Bismuth (Bi) and Element 115 (currently identified as Ununpentium (Uup)) and group 6A has only Livermorium (Lv), which is Element 116. Post-transition metals have metallic appearance (e.g., lustrous and solid at room temperature) and general metallic behavior (e.g., electrical conductivity), but other physical characteristics are different from the metals to their left (alkali, alkaline earth, and transition metals) by being softer and more malleable, and by having lower melting and boiling points.[37] The moniker 'poor metals' has no clear origin, though the 'poor mechanical strength' (referring to the softness of the metal, due to the low melting points and high thermal conductivity) could be part of it, as well as the understanding by chemists that these metals, like the metalloids to their right, show a greater tendency toward nonmetal bonding behavior and crystalline structure than do any of the elements to their left.[38]

The last group of metals is the semimetals, also known as *metalloids*. These members are grouped to the right of the transition metals and include a staggered arrangement of elements, including Boron (B), Silicon (Si), Germanium (Ge), Arsenic (As), Antimony (Sb), Tellurium (Te), and Polonium (Po). These elements are even less metallic than anything found in the transition or inner transition metals groups. But it is this very property that makes these elements critical to many of the electronics we have today.

Metals have such chemical and physical flexibility because the atoms in metals are packed closely together, and electrons that exist in the outermost regions of each atom have the ability to move freely among the atoms in the metal's solid structure. The atoms in metals are considered *delocalized* because of their freedom of movement between and among all the metal atoms in the sample.[39]

This is a good point to look again at the structure of atoms. Metal atoms, like all atoms in our universe, have a dense, massive nucleus of protons and neutrons. Electrons move around the nucleus at fixed distances and in fixed orbits (called *orbitals*). The number of electrons an atom has in its elemental form is equal to the number of protons, so that the overall charge on any elemental atom is zero. And while it seems logical to think that all atoms must exist in their elemental state because the charge on the atom is so lovely, we all know from experience that many elements are found in compounds rather than in elemental forms.

One model used to elucidate atomic structure and bonding behavior (leading to an understanding of the physical and chemical properties used to identify substances) is called the *Bohr model of the atom*. Named for Niels Bohr (1885–1962), its creator, the Bohr model is actually a two-dimensional representation of an atom meant to help us better understand the real three-dimensional

[37] "Chemistry for Kids: Elements - Post-transition, Poor, Other Metals." *Ducksters*. Technological Solutions, Inc. (TSI), Apr. 2015. Web. 17 Apr. 2015. http://www.ducksters.com/science/chemistry/post-transition_metals.php
[38] "Post-transition metal" (2015). Wikipedia.org. Needs correct web address for entry. Accessed April 17, 2015.
[39] Myers, Richard. The Basics of Chemistry (2003). Greenwood Press: Westport, CT. p.80.

atoms that make up matter in our universe. Bohr, a Danish physicist, created the model in his studies of the behavior of elements in the gas phase. He could see that electrons could move and was trying to understand the structure of atoms that gave rise to his observations of his gaseous samples.[40] In the Bohr model, the nucleus of an atom is identified sparsely, with only basic information about protons and neutrons. The most important part of the model is in the concentric rings, which surround this minimal nucleus. Bohr's work made more clear that while electrons do exist in a 'cloud'-like region swirling around the nucleus of an atom, that those same electrons exist only at specific distances from that same nucleus. Each of these distances represents the energy required for the electron to exist at that specific distance from the nucleus. These specific energy levels, called *quantum energy levels*, are what they are because of the forces required in attraction between the positively charged nucleus and the negatively charged electrons must be balanced by the forces of motion keeping the electrons in orbits around the nucleus. There are specific energies required in an atom to keep all these forces balanced and keep the atoms intact.[41] Too little energy for the electrons and it would slow down and be drawn in to crash into the nucleus because of the magnitude of the attractive force. Too much energy, and all the electrons could spin away completely, leaving just a nucleus with nothing to bind it to other atoms.

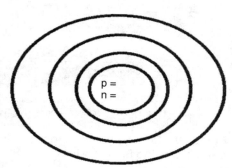

Figure 29 Bohr model template, created by the author.

The organization of electrons in a Bohr model goes from the innermost ring to the outermost and follows a specific pattern that is mirrored in the organization of the periodic table's rows of elements.

Period of Periodic Table	Number of Elements in that Period	Orbit in Bohr Model	Number of Electrons in that Orbit
1	2	Inner most	2
2	8	Second	8
3	8*	Third	8*
4	18	Fourth	18
etc			

*The number of elements that exist in the third period appears to be 8, though modern theories also sometimes include the transition elements from period 4 in the third orbit because studies of the energies of the elements in the fourth period's transition elements indicate these elements are more similar to the energies of the elements in the third period, despite the fact that the atoms of the transition elements of the fourth period are much larger. We will not be studying Bohr models of the transition elements; therefore, we will retain the pattern most obviously indicated by the present structure of the periodic table.

[40] Niels Bohr. (2015). The Biography.com website. Retrieved 09:45, Apr 13, 2015, from http://www.biography.com/people/niels-bohr-21010897

[41] Jackson, Tom (2012). <u>The Elements: An Illustrated History of the Periodic Table</u>. Shelter Harbor Press, New York, p. 92.

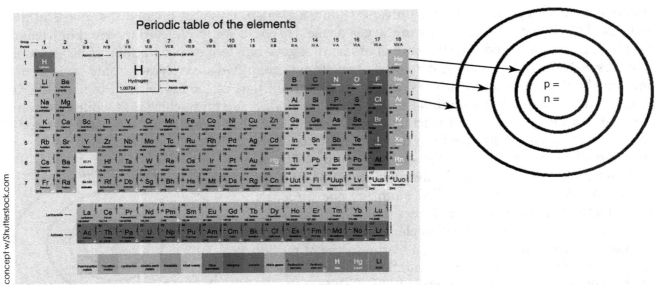

concept w/Shutterstock.com

Figure 30 Comparison of arrangement of elements in the periodic table and the arrangement of electrons in the orbits of the Bohr model.

Let's look at the Bohr models for atoms of some metallic elements. Look for similarities in electron arrangement between atoms in the same group.

Element	Number of Electrons in Atom	Bohr Model of Electron Arrangement
Li	3	p = 3 n = 4
Be	4	p = 4 n = 5
Na	11	p = 11 n = 12

Element	Number of Electrons in Atom	Bohr Model of Electron Arrangement
Mg	12	p = 12 n = 12
Al	13	p = 13 n = 14
K	19	p = 19 n = 20
Ca	20	p = 20 n = 20

In metals that are classically good conductors of electricity, atoms have electrons that are free to move; specifically, these atoms have just one or two electrons in the outermost shell, which is called the *valence shell*. Valence shell electrons are known as *valence electrons* (big surprise!). We can see from the Bohr models that there is lots of room in the valence shells of atoms. This room allows the movement that we interpret on a macro scale as *electricity* and which, as a physical property, we call *electrical conductivity*.[42] Similarly, this movement of electrons can allow electrons to be lost as well, during the chemical reactions with metals that create compounds. This is a property known as *ionization potential*. Metal atoms are known to lose valence electrons easily, and this is because having those few electrons hanging around in the valence shell is more work for an atom than to have all shells filled

---

[42] Gertner, Jon. <u>The Idea Factory: Bell Labs and the Great Age of American Innovation</u> (2012). Penguin Press: New York, NY. p. 83.

to capacity with electrons. Another way to understand this situation is to say that an atom is more energetically stable to have completely filled shells, a completely empty valence shell, or, as a distant third option, a valence shell that is exactly half full. The loss of an electron or electrons from a metal atom in order to bring that atom closer to a state of atomic Nirvana creates a positively charged *ion*, specifically called a *cation*. Cations are part of the chemistry that comprises all salts.

But metals are not the only parts of salts. Metal atoms must chemically combine with nonmetal atoms. Nonmetal elements are what's left after identifying all the metals and metalloid elements. They comprise only a handful of elements out of the 118 currently presented on the periodic table.

Region	Member Element	Common Name
Elements of groups 4A, 5A, and 6A	Carbon (C), Nitrogen (N), Oxygen (O), Phosphorous (P), Sulfur (S), and Selenium (Se)	Other nonmetals
Group 7A	Hydorgen (H)*, Fluorine (F), Chlorine (Cl), Bromine (Br), Iodine (I), Astatine (At), and Element 117 (Ununseptium (Uus))	Halogens or halides
Group 8A	Helium (He), Neon (Ne), Argon (Ar), Krypton (Kr), Xenon (Xe), Radon (Rn), and Element 118 (Ununoctium (Uuo))	Noble gases

Nonmetals are often identified by the characteristics that are decidedly nonmetallic (makes sense). Nonmetals are often (though not exclusively) not solids at room temperature, and if they are, those solids are dull in appearance, brittle, and have relatively low melting and boiling points. Nonmetals do not conduct heat or electricity well (this is sometimes known as being an *insulator*).

Nonmetal chemistry, though, is governed by the state of electron arrangement in shells around the nucleus, which is a characteristic shared by metal atoms as well. And like metals, nonmetals love to have completely filled or completely empty valence shells. But unlike metals, nonmetals atoms are driven toward atomic Nirvana by *gaining electrons*, rather than losing them. Atoms that gain electrons are called *anions*. Bohr modeling can help show why nonmetals crave electrons and how many. As you look at the table below, ask yourself how many electrons at atom of the given element would want in order to fill its valence shell the rest of the way.

Element	Number of Electrons in Atom	Bohr Model of Atom
C	6	
N	7	

Element	Number of Electrons in Atom	Bohr Model of Atom
O	8	p = 8 n = 8
F	9	p = 9 n = 10
Ne	10	p = 10 n = 10
P	15	p = 15 n = 16
S	16	p = 16 n = 16

Element	Number of Electrons in Atom	Bohr Model of Atom
Cl	17	p = 17 n = 18
Ar	18	p = 18 n = 22

The improvement in atomic stability experienced by each atom in a salt compound through losing or gaining electrons drives the chemical likelihood of the chemistry happening to form that compound. Salts are notoriously stable compounds because the ions created during the chemical reaction have electron arrangements that are more desired than those of the atoms they evolved from. In addition, we will see that the ion charges created result in compounds with overall charges of 0, which is also a driving factor in compound stability.

So salts are composed of metal cations with nonmetal anions. Formulas of simple salts indicate this. For example, consider these *binary salts* (salts made of only two elements):

$NaCl$	Sodium chloride (table salt)
$MgBr_2$	Magnesium bromide
CaF_2	Calcium fluoride
$AlBr_3$	Aluminum bromide
Li_2O	Lithium oxide
Fe_2O_3	Iron (III) oxide

In each salt's formula, only two elemental symbols are presented: one is a metal element and the other is a nonmetal element. The subscripts in the formulas indicate the number of atoms of each element, which are required to make a stable salt compound. Binary salt names are also relatively straightforward. The metal element name is written in full, and the nonmetal element name is truncated before the second vowel (don't forget that the y in oxygen is considered a vowel in this case) and the ending 'ide' is put on. How can we know whether the formula for a binary salt is correct? We use the understanding of atomic structure presented in the Bohr models, as well as remember that the charges on all ions in a stable binary salt will add up to zero.

For example, consider the formula for table salt, NaCl. According to the Bohr model for each element in this compound, sodium has one lone electron in it's valence shell, which is one more than having a lovely set of completely filled shells. A sodium atom will look for a chemical reaction where it can give away it's lone electron, resulting in a sodium ion with a +1 charge. In a complementary way, chlorine atoms contain 17 electrons with an arrangement having seven electrons in the valence shell. The chlorine atom is just one electron short of having its valence shell (along with all the other inner shells) completely filled. The chlorine atom is driven to gather another electron if at all possible. This desire for additional elec-

Figure 31 Movement of electron from valence shell of Na to valence shell of Cl.

trons in order to obtain a more favorable energy state is called *electron affinity*. As you might expect, electron affinity increases as we move from left to right across the elements of a period and is strongest among nonmetals. In salt compounds, the reaction forming them has enabled the sodium atom to lose it's electron and for the chlorine atom to gain it. This is a win–win situation, resulting in ions with filled valence shells and with charges that are equal in magnitude and opposite in charge, giving a net overall charge of 0 to the salt compound.

What kind of chemical reactions result in the formation of salts? There are a few that work reliably, but the most common example of a chemical reaction that creates salt products is called a *neutralization reaction*. This is also known as an *acid-base reaction*. In it, an acidic compound reacts with a basic compound to create a salt and a molecule of water. Acidic compounds are those compounds whose formulas carry a hydrogen out front, and generally fall into two classes: binary acids and oxyacids. Look at the table below and see what you can identify as true about these kinds of acids.

Binary Acids	
HF	Hydrofluoric acid
HCl	Hydrochloric acid
HBr	Hydrobromic acid
HI	Hydroiodic acid
Oxyacids	
HNO_3	Nitric acid
HNO_2	Nitrous acid
H_2SO_4	Sulfuric acid
H_2SO_3	Sulfurous acid

Binary acids contain only two elements: hydrogen and another nonmetal element, usually a halogen from Group 7A. Binary acids are named with the prefix *hydro*, followed by a truncation of the other nonmetal element name, and the –*ic* suffix. The word *acid* is written second. Oxyacids contain hydrogen, oxygen, and another nonmetal element. There are many more oxyacids than are presented in this table. Oxyacid names do not use the *hydro* prefix, but instead it start with the truncated version of the nonmetal that is not hydrogen or oxygen. The absence of the hydro prefix already communicates that this is an oxyacid compound, and not a binary acid. Only the other non-metal element (i.e., not hydrogen or oxygen) is directly indicated in the name of the acid. However, the suffix added to the nonmetal name gives information about the number of oxygens in the oxyacid. This is somewhat of a memorize-and-regurgitate situation, as there is not an obvious pattern to the suffix choice and the number of oxygens except to say that in oxyacids where more than one combination of the same three elements exist, the arrangement with more oxygens gets the –*ic* suffix and the one with fewer oxygens gets the -*ous* suffix. As we come across other oxyacids throughout the course, we can add them to the list we have started here.

Bases, on the other hand, are not acids. The bases involved in most acid–base reactions that form salts are hydroxide bases, meaning that they contain the −OH (hydroxide) functional group. The other part of the base is a metal element. Here are some examples of common hydroxide bases:

Hydroxide Bases	
LiOH	Lithium hydroxide
NaOH	Sodium hydroxide
$Mg(OH)_2$	Magnesium hydroxide
KOH	Potassium hydroxide
$Ca(OH)_2$	Calcium hydroxide
$Al(OH)_3$	Aluminum hydroxide

There are many hydroxide bases. The hydroxide group combines easily with many different metal ions to form this kind of base. You should be able to see that naming hydroxide bases is much easier than naming acids. It is simply the metal element name, with the *hydroxide* after that. Nothing else is needed.

Okay, the point of all this is to discuss salts, not acids and bases, but these are all intimately connected in neutralization reactions. Here is an example of a neutralization reaction:

$$HCl + NaOH \rightarrow NaCl + H_2O$$

$$acid + base \rightarrow salt + water$$

All acids reacting with hydroxide bases will proceed to some degree to form a salt compound and a molecule of water. If you look closely, you should be able to identify how the rearrangement works:

$$HCl + NaOH \rightarrow NaCl + H_2O$$

The acid compound donates a hydrogen atom to form water and the chlorine atom to become the anion part of the salt. The base donates its hydroxide part to help form the other part of the water molecule and the sodium metal cation becomes the partner to the chloride anion. The water can be the hardest to visualize:

$$H^+ + OH^- \rightarrow HOH \; H_2O$$

See how the H and OH become water (H_2O)?

What salt do you think would form from a neutralization reaction between LiOH (lithium hydroxide) and HCl (hydrochloric acid)?

$$HCl + LiOH \rightarrow ? + H_2O \quad \text{[Answer: LiCl]}$$

What about a situation that is a little more complex: a neutralization between HCl and $Mg(OH)_2$?

$$HCl + Mg(OH)_2 \rightarrow ? + H_2O$$

You might understand intuitively that the cation (Mg) and anion (Cl) come together to make the salt. But is the formula MgCl or something else? Go back to the Bohr models and figure out what kind of cation and anion are formed. The charges on the ions will help direct the understanding of the salt, which would be created in this neutralization reaction, remembering that 1) ions form so that valence shells will be completely filled or completely emptied and 2) salts form in such a way that the ion charges add up to zero.

Atom	Bohr Model of Atom	Ion Formed	Bohr Model of Ion
Mg	$p = 12$ $n = 12$	Mg^{2+}	$p = 12$ $n = 12$
Cl	$p = 17$ $n = 18$	Cl^-	$p = 17$ $n = 18$

You might be saying to yourself *Sure, I can see why each atom loses or gains the electrons it does because the end result in the ions is empty (in the case of magnesium) or filled (in the case of chlorine) valence shells. But the number of electrons magnesium loses (2) is not the same number of electrons than an atom of chlorine needs (1). What does this mean for the formation of the salt?* This is a great thought. You are absolutely right that sometimes the number of electrons lost by the metal atom is not neatly the exact number of electrons needed by the nonmetal atom. That's okay, but these reactions are not happening one at a time; there are thousands of molecules of acids and bases reacting, and they react in a way that perfectly balances electrons lost and gained. In our example here, each magnesium ion has lost two electrons, and this means that two chlorine ions can each have one. The end result is a salt with the formula $MgCl_2$. This is correct because the charges on all the ions involved adds up to zero.

Other chemical reactions create salts as well. Metals can react with acids to form salts and hydrogen gas. This is a classic gas-forming reaction. For example:

$$Mg(s) + H_2SO_4(aq) \rightarrow MgSO_4(aq) + H_2(g)*$$

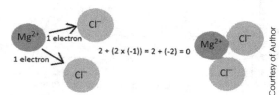

Figure 32 Creation of $MgCl_2$. Confirmation that all charges add up to zero.

Figure 33 Picture of a gas-forming reaction.

(*Remember that chemical reaction equations are most complete with information about the physical state of every component. (s) means solid; (aq) means dissolved in water; (g) means gaseous; and (l) or (ℓ) means a pure liquid not dissolved in water.

Alternatively, salts can be formed by direct synthesis from their elemental components. These are called *synthesis reactions*. Consider this example:

$$Ca(s) + Cl_2(g) \rightarrow CaCl_2(s)$$

Figure 34 Pictures of elements and compound for CaCl₂ synthesis reaction.

As we have learned, the formation of salts happens often when the energy of the reaction is favorable. This includes an understanding of the beneficial loss or gain of electrons to form ions whose charges complement one another in the final products. Because a picture can worth a thousand words, there are diagrams chemists often use to describe the energy profile of a chemical reaction. These are called *energy diagrams*. How original.

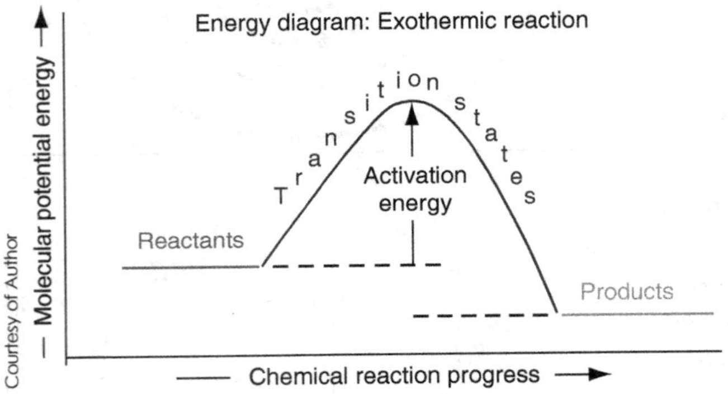

Energy diagrams give information about the amount of energy in the reactants and products (and their intermediates) at every stage of a chemical reaction. From an observer's point of view, the energy of a reaction is often in the form of heat, though light is also sometimes observed. If energy is given off over the course of a reaction, it is called an *exothermic reaction*. The energy diagram for an exothermic reaction could look something like this:

Figure 35 Exothermic energy diagram.

Notice most importantly that the energy level for the reactants is higher than that of the products. This indicates that there has been energy lost, and this energy is assumed to have been lost to the environment outside of the reaction system itself. As chemists, we could monitor the amount of heat energy lost by the system by measuring the temperature of the reaction system with a thermometer. A rise in the temperature would indicate the energy that was once in the molecules of the reactants is more than the energy needed to keep the products together. The excess energy is measured as a rise in the temperature. Many neutralization reactions are exothermic.

The energy diagram also has a center area, which is labeled as *transition states*. This represents the portion of the chemical reaction where the atoms and ions are moving around – where the chemistry is really happening. Notice that the energy of the transition state region is higher than that of the reactants and of the products themselves, and that there is a peak in this region. This can be more easily understood using the analogy of a roller coaster. Roller coaster rides always begin with a slow ascent up an incline so that at the peak of that incline the coaster will rapidly descend and have enough energy in forward momentum to keep you moving through the next series of hair-raising twists, turns, and loops.

In the same way, a chemical reaction begins when the reactants' atoms amass enough energy to begin separating from each other and begin forming new products. The energy required to accomplish the transitions from reactants to products is called the *activation energy*, and each chemical reaction has a unique value for the activation energy required.

One great application for an exothermic reaction is the one harnessed by the military in MREs (meals ready to eat). These are self-contained, nutritionally complete meals that use a reaction between powdered magnesium and water to create enough heat to cook the food in the pouches. This is the chemical reaction equation for the heat source in an MRE:

$$Mg(s) + 2H_2O(l) \rightarrow Mg(OH)_2(aq) + H_2(g) + energy[43]$$

Figure 36

You might notice the similarity of this reaction to the type of gas-forming reaction, which could form a salt. These reactions are very similar, except that in the MRE reaction, the 'acid' is water, and, rather than creating a salt, this reaction produces the basic compound magnesium hydroxide.

Not all reactions are exothermic. Some actually need more energy to be installed in the products than the reactants have available in their original molecules. In a situation like this, the reaction system must absorb energy from the environment, and we can measure this as a drop in temperature over the course of the reaction. A reaction system that results in absorption of energy (a drop in temperature) is called an *endothermic reaction*. Here is a general energy diagram for an endothermic reaction system:

What is the most obvious difference in the endothermic energy diagram compared to one for an exothermic system? Look at the energy levels of the reactants and products. The products exist at a higher energy level than the reactants. Many endothermic reactions are not as spontaneous as most exothermic ones. One example of an endothermic chemical reaction is that of photosynthesis, which we know is the process by which plants produce sugar. The energy provided in the

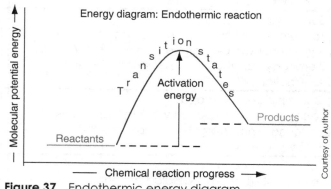

Figure 37 Endothermic energy diagram.

sun's rays makes possible the rearrangement of carbon dioxide gas and water molecules into simple sugars and oxygen gas. The chlorophyll compound in plants absorbs the energy from the sun (a little like solar panels absorb the sun's energy) and then adds that environmental energy to the chemistry of sugar formation:

$$Energy + 6CO_2(g) + 6H_2O(l) \rightarrow C_6H_{12}O_6(aq) + 6O_2(g)$$

Finally, let's touch on how salt (sodium chloride) is produced commercially. While it is possible to create salt chemically through neutralization, gas-forming and/or synthesis reactions, it is usually more practical to harvest the salt that Mother Nature has already made. There is salt dissolved in the

---

[43] Torres, James and Lee, Jae (2000) "Feeding the Army Using a Flameless Ration Heater". Available at http://chem200.tripod.com/. Accessed on April 18, 2015.

water of the oceans and in groundwater, and there are also a few places where ancient oceans dried up and left solid salt behind in the earth crust.

Salt that is dissolved in ocean water can be separated from the water through the simple process of *evaporation*. Evaporation is a lot like boiling, where a material passes from the liquid state to a gaseous state, but evaporation is slow, usually accomplished by allowing the sun to dry shallow pools of seawater. Evaporation is an endothermic physical change process (so is boiling, for that matter) that capitalizes on the fact that the amount of energy needed for a water molecule to transform from liquid to gas is less than the energy required for a solid salt molecule to melt from solid to liquid or even pass from liquid to gas. Salt's molecular stability gives it a lot of internal energy that protects it from molecular transformations of any kind unless there is a lot of energy provided – more energy than the sun can provide simply by shining down on a solution of water and salt. We already know that solutions are mixtures of one substance physically dissolved into another, and that mixtures can be separated by physical means.[44] This means we can apply techniques like those listed below:

Physical Separation Technique	Description
Filtration (picture of colander and draining pasta)	Using a physical, porous barrier (like filter paper or screens of different sizes) to separate materials of different sizes or different physical states from one another
Distillation/condensation (picture of biodiesel synthesis. Example can be found in http://www.chemistryland.com/CHM151W/12-Final/DistillationDiagram.jpg)	Using differences in boiling points to separate materials that may have the same physical state (e.g., liquid) from each other
Sublimation (picture of purification of iodine from a mixture with sand. Example can be found in http://www.benwiggy.com/homework/science/chemistry/sublimation.gif)	Using the ability of some materials to pass directly from the solid phase to the gas phase (sublimation) to separate mixtures of solids
Column chromatography (picture of chromatography column. Examples can be found in http://orgchem.colorado.edu/Technique/Procedures/Columnchrom/Procedure.html)	Using the molecular attractive forces of a filtration medium (often sand, SiO_2) and certain materials to separate mixtures
Solubility/extraction (picture of oil-and-water salad dressing)	Using the ability of some materials to dissolve in some solvents better than in others
Centrifugation (picture of before and after of contents of a centrifuge tube. Example at https://s.yimg.com/fz/api/res/1.2/dDibitZn9CW.jYO.N4CUpQ—/YXBwaWQ9c3J-jaGRkO2g9MjY0O3E9OTU7dz0zMDA-/http://upload.wikimedia.org/wikipedia/commons/thumb/4/49/Image1centri.JPG/300px-Image1centri.JPG. It's in French, but it should be pretty obvious what is going on)	Separating solids and liquids by spinning the mixture in a centrifuge, which uses centrifugal force (outward forces created by circular motion) to concentrate the substance with greater density away from the substance(s) with lower density

Sea salt is harvested often through solar evaporation, which capitalizes on the fact that salt is soluble (dissolves) in water, and that the boiling point of water is much lower than the melting and boiling points of salt. Solar evaporation facilities will pump seawater into large, shallow pools and then let

---

[44] "Separation Techniques" (n.d.). Available at http://www.kentchemistry.com/links/Matter/separation.htm. Accessed on April 19, 2015

the sun slowly 'boil' the water off through evaporation, leaving behind the solid salt. Seawater often contains more salts than sodium chloride (NaCl). There can also be calcium sulfate ($CaSO_4$, also called *gypsum*), magnesium chloride ($MgCl_2$), potassium chloride (KCl), and magnesium sulfate ($MgSO_4$).[45] However, sea salt refining processes can separate the NaCl from the other salts very reliably by capitalizing on the fact that sodium chloride is pretty

Figure 38 Picture of solar salt refining process.

soluble in water until the concentration of salt reaches a concentration of about 25% by weight[46] in the concentrating pool. (Seawater is 3–4% salinity – salt content – when it is in the ocean. This translates to about 35 g of salt in every liter of water.)[47] Once the concentration of salt reaches this point (because water has been lost by evaporation), the salt will no longer be as soluble in the water and the salt will solidify as crystalline salt. This situation can then use filtration to separate the solid salt from the rest of the water and the salts that are still dissolved in it, or simply rake up the solid salt and gather it into baskets for further drying and packaging. Processes that rely only on solar evaporation can take up to five years to get to the point where NaCl will crystallize and can be harvested.

In a similar fashion, salt can be harvested far from the sea if a *brine spring* is nearby. A brine spring is a place where fresh water passes through layers of mineral-rich rock on its way underground. Through its motion in the earth's crust, the water dissolves many salts along the way, one of them being sodium chloride. Once brine springs are located, the basic process of harvesting salt (known as 'halite' in some literature on brine spring salt production)[48] is similar to sea salt production, in that the point of refining is to separate the salt from the water and other mineral compounds, which might also be dissolved. Brine salt refining often pumps the water into large shallow pans, which are heated in order to speed the process of evaporation of water from the salt solution. Once the concentration of salt has reached the point where salt crystallization begins, the salt can be harvested by gathering the solid salt out of the rest of the brine, washing it and packaging it for use.

There is one other major way of collecting salt: mining it! Solid salt, often called *rock salt*, is found in places in the earth's crust where ancient seas once sat but evaporated, leaving behind all the salt and other minerals that had been dissolved in the water. Rock salt mining is similar to other kinds of mining, in that a hole in the ground is created, and the salt is dug out and carried to the

---

[45] Wilkins, Guy (n.d.) "Solar Salt Engineering" Available at http://solarsaltharvesters.com/notes.htm. Accessed on April 19, 2015.

[46] The concentration or density of salt water is measured in older literature as 'degrees Baumé', abbreviated 'o Be' or 'deg Be'. It is a straightforward conversion of units: 25% by weight in water is the same as 25° Be. [*Webster's Revised Unabridged Dictionary*. S.v. "Baume." Retrieved April 19 2015 from http://www.thefree-dictionary.com/Baume].

[47] Christensen, Emma (2011) "From Ocean to Box: How Sea Salt is Harvested". Available at http://www.thekitchn.com/come-along-on-a-159478. Accessed on April 19, 2015.

[48] "Salt Production in Syracuse, New York ('The Salt City') and the Hydrogeology of the Onondaga Creek Valley". Publication of the US Geological Survey, Fact Sheet FS 139-00, November 2000. Available at pubs.usgs.gov/fs/2000/0139/report.pdf. Accessed on April 19, 2015.

surface for purification. Salt mining often uses the *room and pillar* mining method, which leaves pillars of salt to support the roof of the mine. This method makes it possible to safely excavate 50–65% of the rock salt in the mine, which making it unnecessary to construct much additional supporting structures.[49] Mining rock salt has some advantages to other mineral excavation, in that while most of the time, temperature in a mine increases as the mine goes deeper toward the earth's core; in rock salt mines, the temperature is stable, around 70 °F all the time![50] Rock salt is first crushed and then removed from the mine by conveyor belts end elevators to the surface where it is then sorted for size, packaged, and shipped for sale around the world.

Figure 39   Picture of salt mining.

Each of these salt collection options has its advantages and disadvantages, but what we absolutely know is: we have to have salt because our very lives depend on it. Why do we need salt? There are many small specific reasons for salt in the human diet, but they all fall into three main classes of need: for proper nerve impulse transmission; for successful transportation of materials into, out of, and between cells; and for maintaining proper moisture (water) balance in all the cells of the body.

Nerve impulses occur when sodium and potassium ions gather in unequal concentrations on either side of a nerve cell membrane. The different concentrations build up a charge differential across the membrane; think of this is as similar to the kind of charges that build up in the atmosphere in advance of a lightning bolt discharge during a summer storm. When the charge differential is great enough, the nerve cell opens up special channels in the cell membrane and the ions flow in and out in directions, which aim to restore electrical balance to the cell. This ion movement discharges the electrical differential, and the result is a transfer of an electrical impulse. This system of ion movements and electrical discharges is called the *sodium–potassium pump*.[51,52,53]

Muscle cells also utilize the sodium–potassium pump in order to drive muscle contractions.

Sodium ions from salt also drive several *indirect pumps*, which result in the movement of crucial materials in and out of cells through membranes. For example, there is a situation where sodium ions and glucose can enter a cell together through the action of indirect pumps called *symport pumps*. Similarly,

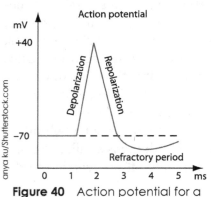

Figure 40   Action potential for a nerve cell.

[49] "Salt" (n.d.) Available at http://earthsci.org/mineral/mindep/salt/salt.htm. Accessed April 19, 2015.

[50] "Salt Production and Processing" (n.d.) Morton Salt Company. Available at http://www.mortonsalt.com/salt-facts/salt-production-and-processing. Accessed April 19, 2015.

[51] "The Sodium-Potassium Pump" (n.d.) Available at http://hyperphysics.phy-astr.gsu.edu/hbase/Biology/nakpump.html. Accessed on April 19, 2015.

[52] "Action Potential" (n.d.) Available at http://hyperphysics.phy-astr.gsu.edu/hbase/Biology/actpot.html#c1. Accessed April 19, 2015.

[53] Chudler, Eric. (2010) "Lights, Camera, Action Potential!" Available at http://faculty.washington.edu/chudler/ap.html. Accessed on April 20, 2015.

symport pumps move amino acids in and out of tissues; they return neurotransmitter molecules (molecules that aid in the electrical nerve impulse as it moves from one nerve cell to another) back to their original nerve cells in preparation for new nerve impulse transmission needs; and pump iodine ions into the thyroid gland in order to help that gland create the hormones needed to regulate metabolism, growth, hearth rate, temperature, and more.[54,55]

Moisture balance is also maintained in part by salt, by the attraction of water molecules to the charged ions that make up salt. In animal cells, water tends to flow from areas of higher concentration to areas of lower concentration – a process called *osmosis*.[54] This osmosis can be enabled by differences in concentrations of salt in a system. When greater concentrations of sodium ions exist outside of a cell, water is drawn out of the cell and the result is a dilution of the extracellular fluid. In this way, the body manages its moisture balance at the cellular level. On a larger scale, we also move water around using salt when we need to cool off. Sweat produced during exercise is done so in order to cool the body. Liquid water on the skin will evaporate using the heat from the skin. This lowers the temperature of the body. In addition, salt is excreted in sweat as well so that the blood's salt level would not become too high when water is lost through sweating.[56]

And while it seems obvious at this point that salt is crucial for proper life processes, it is also possible to have too much or too little salt in the diet. These conditions result in health problems that can be deadly. Diets which contain too little salt can lead to increases in blood lipid levels (dissolved fats in the blood) and cholesterol, which in turn increases the risk of heart disease and problems like stroke and heart attack. Development of type 2 diabetes (sometimes referred to as adult-onset diabetes) is also connected to diets with too little salt, and there is evidence that those with type 2 diabetes tend to die earlier when their diets include too little salt. Finally, some studies indicate that declines in cognitive ability and increases in the risk of falls are connected to *hyponatremia*, a condition where there is too little salt in the body.[57] Conversely, too much salt can also lead to higher risk of heart attack and high blood pressure from the inability of the kidneys to keep up with the removal of the excess fluids present when too much salt is present in the blood.[58]

---

[54] "Function of Iodine: Why Do I Need Iodine?"(2011). Available at http://www.saywhydoi.com/function-of-iodine-why-do-i-need-iodine/. Accessed on April 20, 2015.

[55] "Transport Across Cell Membranes" (2014). Available at http://users.rcn.com/jkimball.ma.ultranet/BiologyPages/D/Diffusion.html. Accessed on April 20, 2015.

[56] Roberts, Thom (n.d.). "Why is Perspiration Salty?" Available at http://www.ehow.com/facts_7461998_perspiration-salty_.html. Accessed April 20, 2015.

[57] "Salt is an Important ingredient in your good health. . ." (2015). The Salt Institute. From a fact sheet available at http://www.saltinstitute.org/health/overview/. Accessed April 20, 2015.

[58] "Health Risks and Disease" (2015). Harvard T.H.Chan School of Public Health. Available at http://www.hsph.harvard.edu/nutritionsource/salt-and-sodium/sodium-health-risks-and-disease/. Accessed on April 20, 2015.

SPICES

As we think about food, we can't avoid talking about more than just sweetness, saltiness, or whether starch or protein was the main course. Without flavor, food would be much less enjoyable, though perhaps no less nutritious. Fortunately, many foods also contain compounds that make them enjoyable to the taste.

Spices are usually understood to be plant material other than leaves and are generally dried.[59] Herbs are the leaves of a plant, either fresh or dried. Herbs and spices provide interest to the tongue and bring out more subtle flavors in foods or augment a specific texture of a given food. The biological benefits of spicy foods include a cooling effect in hot climates and some antimicrobial action to help treat and prevent disease.[60] But it's the chemistry of these plant products that provides these properties, taste as well as function.

Chemists have come to classify many of the compounds in spices as *alkaloids*, named from the Arabic *al-qali*. Many of these compounds have a chemically 'basic' nature; this does not mean simple, but represents chemical behavior that is the complement to acidic behavior. In human experience, basic compounds are often bitter tasting, and sometimes have a kind of soapy feeling on the skin. Plants that are harvested for spices contain these spicy compounds for reasons other than to make our suppers more interesting. These compounds are made as part of an arsenal of protective chemistry for the plant. Much of the time these compounds are actually meant to be poisons to the predators of the plant. They are designed by the plant to have a negative biological effect on any creature unwise enough to try to eat it. The goal of any plant is to preserve itself long enough to reproduce and to protect its offspring (seeds). Alkaloid compounds aid in this primal process.

Alkaloids, in their earliest chemical definition, are organic compounds, which contain nitrogen, often (but not always) in some kind of ring structure in the molecule.[61] These molecules also cause a biological response in organisms that ingest them, which pharmaceuticals both ancient and modern have been capitalizing on. Opium, tobacco, coffee, chocolate, chili peppers, and many other substances contain alkaloids, which give the end products properties that have a market in today's world.

---

[59] http://www.spiceislands.com/SpiceEducation/HerbsVsSpices.aspx accessed March 23, 2015.

[60] Pollan, Michael (2008). In Defense of Food. Penguin Group (USA): New York, NY. pp. 173–174.

[61] Robinson, Trevor. "Metabolism and Function of Alkaloids in Plants" Published in *Science*, New Series, Vol. 182, No. 4135 (April 26, 1974), pp. 430–435. Obtained via http://www.jstor.org/stable/1738505. Retrieved March 23, 2015.

Most people, when they hear the word 'spicy', think of the sensation of *HOT*. Foods that taste hot are so because the compounds in them interact with receptors in the mouth, which are programmed to sense pain. As in 'Ow! That's hot!' Even if spicy food is consumed cold, the hot sensation is still there. This is because it is a chemical interaction, not a thermal one. Like the taste buds that sense sweet compounds, there are structures in the mouth (and elsewhere) called *nociceptors* that are designed to sense pain, cold, pressure, and heat. Nociceptors are nerve endings bundled in the outer layers of skin and in moist tissues like those in the mouth and nose. When a chemical activates a nociceptor and causes a sensation that is interpreted as something else – cold, heat, pain, for example – this situation is called *chemesthesis*. Spicy alkaloids do just that.

The partner to salt on most tables is black pepper, or just *pepper*. Black pepper has a spicy flavor in the mouth, and even stimulates sneezing if it gets up into nasal tissues. What is it about pepper that gives us this kind of biological response? What message is the spiciness and the sneezing communicating? And why do we *like it*?

The spice we identify as black pepper is usually ground black peppercorns, which are actually the dried, fermented seeds of the *Piper nigrum* plant, a vine origi-

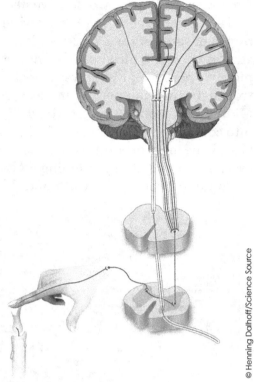

Figure 41 Very basic anatomy of a nociceptor.

nating in the southern tip of India. Calicut, India, is considered to be the birthplace of black pepper, and it was from this area that black pepper started on its journey around the world with Indian and Arab traders. From Calicut, it eventually spread to Europe by way of Venice, which was *THE* destination point for exotic products that had European demand. Venice was an end-point trading post, which grew in prominence especially during the Middle Ages (11th–15th centuries AD) and especially so during the period of the Crusades (1096–1270 AD). As a popular cause, the Crusades were aimed at pushing back Arab (often Islamic) possession of lands which bordered or were in Europe. Arab territorial expansion began with conquest of Jerusalem in the mid-7th century. Controlling the Palestinian region was a central location for spreading north and west into Turkey and the Mediterranean region, which involved numerous clashes over the next 450 years with Byzantine, Turkish, and Roman Empire forces. (Interestingly, the conquest of the Iberian peninsula in the 8th century AD was a blow to Visigoth leaders, descendants of the declining Roman Empire, but was seen as a victory for many in the region for whom stories of the tolerant Islamic forces seemed a better option than religious oppression by the Visigoths.)[62,63] The European (often also labeled Christian) pushback of the Arabian advance began in earnest in the late 11th century with battles reclaiming Jerusalem for the Catholic Church. Soldiers traveling to the Holy Land region

[62] Wilde, Robert "The Visigoths". Accessed on May 23, 2015, from http://europeanhistory.about.com/od/historybypeoples/a/overvisigoths.htm
[63] Alkhateeb, Firas (2013) "Christianity and the Muslim Conquest of Spain". Accessed May 23, 2015; available at http://lostislamichistory.com/christianity-and-the-muslim-conquest-of-spain/

from western Europe would have needed many things for their success, like armor, weapons, and tools, just to name a few. If they were traveling over land (the British and French forces traveled by sea in some of the Crusades), it made sense not to carry *everything* from home, but instead to stop at Venice to pick up what they needed for the next part of the journey. Venice was not completely a part of Italy, but was a nearly independent state ruled by a *doge* (a duke) and which controlled its own commerce. Similarly, soldiers returning from battles in the eastern Europe and the Middle East stopped back through Venice to bring home the exotic products that they had come into contact with in their travels. Black pepper was one such product. Middle Eastern cuisine had already been incorporating black pepper for at least 2000 years by the time the Crusades began[64] as a partner to salt in preserving foods through techniques like brining. Black pepper has the ability to inhibit bacterial growth and this property made it a valuable food preservative product in the age before refrigeration. So valuable, in fact, that in some times and places, peppercorns were used as currency because the coinage of the region varied in precious metal content and therefore it's worth also varied.[65]

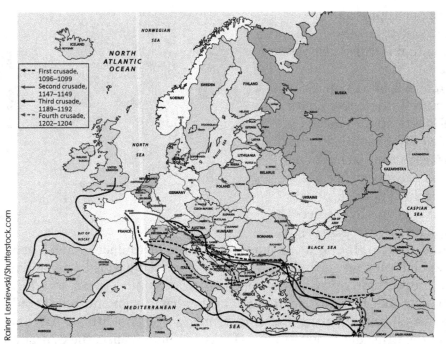

Figure 42 Map of the Great Crusades.

The traders in city-states like Venice knew that value of peppercorns and charged exorbitant prices for it to anyone who came to buy it. Shipping the peppercorns from India through trade routes controlled by Indian and Arab merchants and traders was an expensive, mostly overland enterprise, but Venetian trader prices for the spice more than compensated for whatever they paid out of their own pocket to secure it. Eventually, these prices became so high that even the European elite would not pay them; instead, powers like Portugal and Spain (later Britain, France, and Holland as well) sent ships of sailors and soldiers out from their ports to find a route to India[66] in order to get the peppercorns at the their source, skipping the middle man altogether. These actions sparked the historical period now known as the Age of Discovery and began to exact

[64] Butler, Stephanie (2013) "Off the Spice Rack: The Story of Pepper" Accessed on May 25, 2015 from http://www.history.com/news/hungry-history/off-the-spice-rack-the-story-of-pepper

[65] "Pepper: The King of Spices" (n.d.). Accessed on May 25, 2015 from http://theepicentre.com/pepper-the-king-of-spices/

[66] Kreis, Steven (2002) "The History Guide Lectures on Early Modern European History. Lecture 2: The Age of Discovery" Available at http://www.historyguide.org/earlymod/lecture2c.html

changes on the geographical, political, and cultural face of the rest of the earth. For example, the fact that all countries from Mexico southward through Central and South America speak Spanish or Portuguese is a result of Spanish and Portuguese explorations from Europe, which landed in the New World.

Black pepper's value as a food preservative comes from its active ingredient compound, *piperine*, which is classified as an alkaloid. Examination of its structure reveals some distinct features: 1) an aromatic ring, where six carbons form a cyclical structure and double bonds appear between every other pair of carbons in the ring; 2) a section of the molecule where a carbonyl (a carbon atom doubly bonded to an oxygen atom) is next to a nitrogen atom; 3) a section of the molecule where a chain of carbons exists, in which there are some double bonds as well; and 4) two oxygen atoms on opposite side of the aromatic ring from all the other molecular portions.

The best understanding of why piperine tastes hot comes from its molecular shape and the functional regions of the molecule that are indicated in Figure 4. In the three-dimensional world of life, molecules are not flat or two-dimensional, the way the picture might look. The atoms that make up piperine arrange themselves in space in such a way that nociceptors in the mouth react to their presence. The physical response we feel as a result of the presence of a molecule like piperine is created by a cascade of chemical signals that start in the mouth and end in the brain. The brain interprets the message that the chemical signals are communicating (kind of like a chemical Morse code) and tells the body how to respond. The first response is the burning pain we experience – which is meant to stimulate you to spit out the offending material. It is an evolutionary

Figure 43 Chemical structure of piperine. Indicate with numbers the different functional sections of the molecule, based on the descriptions in the paragraph above.

response aimed at keeping us alive by removing the impetus for eating something that may be a poison. The second response is to stimulate the production of molecules called *endorphins*. Endorphins are produced in the brain, in regions of the central nervous system, and by the pituitary gland. Endorphin production reduces sensations of pain and also increases feelings of euphoria, among other things.[67] We will return to endorphins in later discussion on drugs.

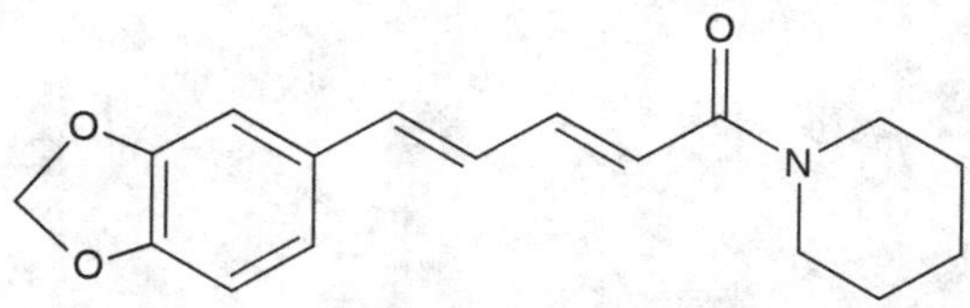

The desire for peppercorns was so strong that Spain's commission of Christopher Columbus charged the Italian explorer to find a westward route to India. What no one knew for sure yet was that there were some continental obstructions between Spain and India: namely, North, Central, and South America. With high hopes, Columbus set sail in August 1492. In October 1492, Columbus landed in the Bahamas and thought he had reached islands off the coast of Japan. By December, he had reached Haiti[68] and there is where the connection to peppercorns changes completely. The diet of Haitian natives (the Taino) included the spicy red or green pods from a plant, which has been in common use among natives in the New World – natives of the American southwest all the way to South America – for thousands of years prior to Columbus' arrival. Perhaps for even as many as 9000 years, indigenous peoples had been incorporating the chili pepper in their food. Beyond diet,

---

[67] Stoppler, Melissa Conrad, and William C. Shiel, Jr. (ed.) (2012) "Endorphins: Natural Pain and Stress Fighters". Accessed from http://www.medicinenet.com/script/main/art.asp?articlekey=55001

[68] *Encyclopædia Britannica Online*, s. v. "Christopher Columbus", accessed May 25, 2015, http://www.britannica.com/EBchecked/topic/127070/Christopher-Columbus/223123/The-first-voyage.

the cultures of the people in the New World also included the chili pepper: it was used to prevent and heal disease, lift depression, to prevent witchcraft, and children who misbehaved were forced to breathe in smoke from burning chili peppers.[69]

Columbus tasted food prepared by local peoples and found it spicier than any European dish made with peppercorns. Lacking any other experience of spices like these, he called the plants *peppers*, though biologically the chili pepper family is not related to that family, which contains black peppercorns. He returned to Spain with seeds of this amazing plant, along with other spoils of his discovery. (Columbus wanted to return with *human cargo*, possible only through kidnapping of local peoples.) And the cuisine of Europe, Asia, and Africa was changed forever. The taste for the spicy pepper spread like wildfire through Asia and Africa first, transforming the dishes there. The mouth-blistering dishes of Taiwan, India, and China, as well as spicy dishes like Hungarian goulash and the spicy sausages and salami of Italy (pepperoni, anyone?) are all what they are because chili peppers arrived on the scene in the early 16th century.[69]

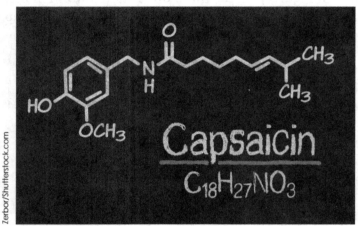

Figure 44   Chemical structure of capsaicin.

So, is there a connection between the peppercorn berry and the fruit of the chili pepper plant? Yes, and no. The chili pepper plant is a completely different species of plant, in the *Capsicum* genus. The pungency of chili peppers comes from the presence of the molecule *capsaicin*:

Notice that capsaicin has some similar chemical composition and structural features to piperine, though the arrangement of these characteristics is different. An aromatic ring is present, which has two oxygen atoms connected directly to it, which are on the opposite side of the ring from a carbonyl next to a nitrogen atom, and a section of the molecule where a chain of carbons exists, in which there are some double bonds as well. These molecular features, coupled with the three-dimensional folded structure of the molecule and its interaction with nociceptors, make capsaicin a very powerful spice, very HOT: so hot that some species of chili pepper can blister the skin, and worse. This reality is capitalized on with personal protective products like pepper spray, which is a solution of capsaicin and capsaicin-derivative compounds (*capsaicinoids*) dissolved in an alcohol solution. The nociceptors in the eyes are especially sensitive to the capsaicinoid compounds, and the result is terrible burning and tearing up of the eyes.

Like piperine, capsaicin stimulates the sensation of burning pain and also the production of pain-reducing endorphins. So, even though the structures of these two compounds are different, the broad end results are the same.

But Columbus wasn't the only explorer searching for peppercorns. In reality, the spices of the Far and Middle East – spices like peppercorns, allspice, ginger, cinnamon, nutmeg, cloves, cumin, mustard and horseradish, turmeric, and herbs like horseradish and wasabi, and garlic – had become desired by European elite as a result of the Crusades. Slowly, these spices made their way into the diet of the average person as well, which made the demand even greater, driving up the price. Venetian traders handled all kinds of spices, and the prices of all of them were an issue for European buyers. The Age of Discovery was launched to increase European access to these delicious and exotic flavors

[69] Timbrook, Jan (2013) "The Natural History of the Chile Peppers", accessed May 25, 2015, http://www.sbnature.org/crc/332.html

at the best price possible. This usually meant sailing to the source of the spice and taking political control of the region. The fact that Columbus found chili peppers was truly blind luck – since he thought he had landed in the Far East and not a brand new continent in the Western Hemisphere – but the end result for the Americas was similar to other regions of the world where products were sourced: indigenous peoples were taken advantage of (to say the least) and the products were brought back to Europe for ravenous consumption. The only difference is that Venice isn't making the money anymore; instead, it is the governments of the countries who controlled the spice sources who are making money.

Let's look at these other spices, starting with ginger, cardamom, turmeric, cumin, cinnamon and mustard, which are ancient ingredients in Indian, east Asian, and Far Eastern cooking.[70,71,72] Each of these spices contain many different compounds, but there are chemicals within them that provide the characteristic taste most generally associated with that spice. The table below summarizes these spicy compounds.

Spice	Active Chemical Ingredient(s)
Ginger	Zingerone.[73,74]
Cardamom	Alpha-terpinyl acetate and 1,8-cineole[75]
Turmeric	Curcumin[76]

[70] "Use of Spices in Ancient India" (2008). Accesed May 27, 2015. Available at http://www.indianetzone.com/52/use_spices_ancient_india.htm

[71] Iyer, Raghavan. *660 Curries*. (New York, NY: Workman Publishing, c.2008), p.2–3.

[72] Rayment, W.J. "The History of Cinnamon" Accessed on May 28, 2015. Available at http://www.indepthinfo.com/cinnamon/history.shtml

[73] LeCouteur, Penny and Jay Burreson (2005) Napoleons Buttons, Penguin Group, New York, p 26.

[74] Image of zingerone from http://www.chemspider.com/ImageView.aspx?id=28952

[75] Cardamom (Elettaria cardamom) (n.d)" Web. Accessed on May 28, 2015. Available at http://www.sigmaaldrich.com/life-science/nutrition-research/learning-center/plant-profiler/elettaria-cardamomum.html

[76] Ingredient in turmeric spice when combined with anti-nausea drug kills cancer cells" (n.d) Web. Accessed on May 28, 2015. Available at http://phys.org/news/2013-08-ingredient-turmeric-spice-combined-anti-nausea.html

Cumin	Nigellone, aka damascenine,[77] and cuminaldehyde[78]
Mustard	Allyl isothiocyanate[79]
Cinnamon	Cinnamaldehyde[80]

Structures from chromatos/Shutterstock.com

What can we notice about these spices? From personal experience, it might be understood that not all have the pain-inducing effect of piperine and capsaicin; but there are some commonalities among these spices that we can see: 1) all are organic; 2) all but one (isothiocyanate, from mustard) have oxygens in the structure, and of these; 3) three have oxygen atoms immediately attached to a benzene ring (reminiscent of piperine and capasaicin); 4) also among the oxygens is a prevalence of carbonyl oxygens (C=O); 5) five are aromatic; 6) two contain nitrogen atoms; 7) all but one have some degree of *conjugation* (carbon to carbon double bonds alternating with single carbon to carbon bonds). So perhaps there is something that chemical structure can tell us about spiciness as a physical characteristic. Let's keep looking.

Serban Bogdan/Shutterstock.com

Figure 45   Map of Malaku.

[77] Isaacs, Tony (2010). "Black cumin seeds provide many wonderful health benefits" Web. Accessed on May 28, 2015. http://www.naturalnews.com/030800_cumin_seeds_health.html
[78] "Major Organic Compounds in Herbs and Spices" (2014). Accessed on May 28, 2015. Available at http://www.compoundchem.com/wp-content/uploads/2014/03/Herbs-Spices-Chemical-Compounds.png
[79] "Mustard (condiment). Wikipedia. Web. Accessed on May 28, 2015. Available at http://en.wikipedia.org/wiki/Mustard_(condiment)
[80] Major Organic Compounds in Herbs and Spices" (2014). Accessed on May 28, 2015. Available at http://www.compoundchem.com/wp-content/uploads/2014/03/Herbs-Spices-Chemical-Compounds.png

Some spices have roots (ha) further than India. In the Indonesian archipelago, there exist hundreds of small islands, which collectively have been known as the Spice Islands, but which are now officially called the *Moluccas*, or *Malaku*.[81]

These islands may be physically small, but they have played a huge role in world history as being the intended destination for larger world powers like Spain, Portugal, Britain, and Holland (and later Germany during World War II). Spices like cloves, nutmeg, and allspice had been traded by Arab traders exclusively until the European palate tasted them near the end of the Crusades. Efforts to politically, militarily, and economically secure the geographical home of these spices were part of the impetus for European powers' setting to sea during the Age of Discovery. Like other spices we have seen, cloves, nutmeg, and allspice offer flavor, some medicinal applications, and some preservative benefits. The table below summarizes these exotic spices.

Spice	Active Chemical Ingredient(s)
Cloves	Eugenol
Nutmeg	Isoeugenol
Allspice	Eugenol and eucalyptol

Courtesy of Author

Examination of these compounds in the same way we looked at those previously presented reveals even more similarity to one another. All are organic, two of three of the compounds are aromatic, and those same two, eugenol and isoeugenol, are isomers of one another and show up in all three spices, and both contain the oxygen atoms directly bonded to the aromatic ring, and some other carbon chain conjugation. Even in allspice, eugenol is the primary essential oil (representing 60–80% of all essential oil content) responsible for the flavor. Eucalyptol is a minor flavor component in allspice, and is not the only one, but it has the highest concentration of the minor compounds.[82] Eucalyptol is presented here mostly for comparison to the other compounds, but still there is an oxygen atom in the compound, which seems to be a very common complement to the carbons also present in almost all of the spices covered to this point.

[81] "Spice Islands (Moluccas): 250 Years of Maps (1521–1760) Accessed on May 28, 2015. Available at http://libweb5.princeton.edu/visual_materials/maps/websites/pacific/spice-islands/spice-islands-maps.html
[82] "Allspice" (2009). Accessed on May 29, 2015 and available at http://www.drugs.com/npp/allspice.html

Other commonalities exist for all these spices as well. All are plant-based, and all have shown some degree of antimicrobial properties (some better than others).[83] Generally speaking, these spicy compounds are aromatic, contain oxygen close to the benzene ring and other regions of conjugation. What is also true is that the plants are making these compounds as part of their chemical arsenals of protection from predators, at least small ones. Some of these compounds are classic alkaloids (organic, plant-based, contain nitrogen (often in a ring)), and all have some degree of biological activity, both for plant predators and in humans (perhaps humans are the ultimate plant predators), though fortunately the biological effect is not as deadly for humans because of the large size and complexity of the human body. The information that has been learned about spices, the chemical composition and structures of the active ingredients that produce the biological responses we have come to expect – heat (even pain), endorphin production, not to mention flavor – inform decisions about possible biological activity in other substances with other chemical components as well as plant (ha) ideas for new compounds which might be synthesized in the laboratory as spice replacements or new flavoring compounds. For example, consider the below plant-based compound:

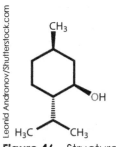

Figure 46 Structure of menthol.

Would this compound be expected to taste spicy, given it's structure? Observation confirms that it is organic, and you have been told that it is plant based, but is it 1) aromatic; does it 2) contain oxygen; and 3) contain regions of conjugation? The answer is 1) no, 2) yes, and 3) no. Therefore, educated guess would support a conclusion that it is not spicy. This compound is menthol, and it tastes cool and minty, not hot and spicy; so the conclusion supports the experience.

[83] Snyder, O. Peter (1997). "Antimicrobial Effects of Spice and Herbs" http://www.hi-tm.com/Documents/Spices.html

POLYMERS AND TEXTILES

A Noiseless, Patient Spider

A noiseless, patient spider,
I mark'd, where, on a little promontory, it stood, isolated;
Mark'd how, to explore the vacant, vast surrounding,
It launch'd forth filament, filament, filament, out of itself;
Ever unreeling them—ever tirelessly speeding them.

And you, O my Soul, where you stand,
Surrounded, surrounded, in measureless oceans of space,
Ceaselessly musing, venturing, throwing,—seeking the spheres, to
connect them;
Till the bridge you will need, be form'd—till the ductile anchor hold;
Till the gossamer thread you fling, catch somewhere, O my Soul

- Walt Whitman

Walk Whitman's musings on the wonder of the spider's thread are no less profound today. Inside the spider's body, a miracle of synthesis occurs nearly instantaneously as the spider weaves a web, lowers itself out of a tree, or wraps its prey. What Walt Whitman didn't know when he wrote this poem was that the filament of the spider's web is a *polymer*, a material made of smaller, usually repeating or similar units bound chemically together to make a larger molecule.

Polymers are all around us, both in nature and in the laboratory. Look around you, starting with the clothes on your back: they are most likely made of cotton, which is a polymer of beta-glucose molecules chemically joined to make a substance called *cellulose*. If you are wearing something with rayon, or nylon, or polyester, those are polymer textiles also. What about in the car you drive? The tires are made of rubber, which is a polymer nature designed but humans have improved. The plastic dashboard, plastic grill pieces, or plastic lenses over the taillights are all made of man-made polymeric materials. The vinyl seat covers are made of polymers. Say you drive to get something to eat. The French fries you eat are an edible polymer called *starch*. The meat in your burger is a polymer of protein. Even inside of the strawberries you might have for dessert is DNA—deoxyribonucleic acid, also sometimes referred to as an organism's *genome*—which is a polymer.[1] We have DNA also; in fact, this DNA dictates everything that our bodies are and do. Without polymers, life as we know it would not exist.

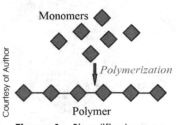

Figure 1 Simplified monomer→ polymer graphic.

The word *polymer* has its roots in Greek: *poly*, meaning 'many', and *meros*, meaning part or portion.[2] So in order for a substance to be classified as a polymer, regardless of where that substance is found or how it is created, it must be substance whose molecules are composed of chains of smaller units, and those units must be identical (starch, cellulose, natural rubber, and many of our man-made plastics and textiles fall into this category), or structurally similar (silk, DNA, and all proteins are examples of this category).

[1] "Deoxyribonucleic Acid (DNA)" (2015). *National Human Genome Project*. Web. http://www.genome.gov/25520880. Accessed May 28, 2015.

[2] McCarthy, E.M. 2013. *Root Word Dictionary*. www.macroevolution.net/root-word-dictionary.html. Accessed May 28, 2015.

POLYSACCHARIDES

Since we have just finished a unit on food chemistry, let's start with some polymers that have connections to food: *polysaccharides*. As the name suggests, polysaccharides are polymers with 'saccharides'—more commonly known as sugars—as the monomer units. The main sugars found in polysaccharides are beta-glucose and alpha-glucose, though by no means are these the only sugar monomers possible?. Polysaccharides fall into two main classes: structural polysaccharides and storage polysaccharides. As might be guessed, *structural* and *storage* refer to the uses of these polysaccharides in nature. The properties of polysaccharides depend on four main factors: 1) polymer length; 2) whether it is a straight chain or branched, and if branched, how much so; 3) how the molecule arranges itself in a three-dimensional way (*folding*) and in a related way; and 4) if it is a chain, whether the chain lies straight in space, or whether it tends to coil up on itself.[3]

Storage polysaccharides are those polymers, which are made by plants and animals to store sugar for later or emergency use. Plants in particular make storage polysaccharides, and most of the time, these compounds are known as *starches*. This makes sense when you remember that green plants make glucose as a result of photosynthesis. And even the most vigorous of plants is not using up all the sugar it is making through photosynthesis. So it is stored by polymerization into polysaccharides. The building blocks of any storage polysaccharide are alpha-glucose, and these monomers are joined chemically together through the condensation reaction, which we saw in Unit 1 is responsible for the formation of sucrose from alpha-glucose and beta-fructose. The formation of

Figure 2 Condensation reaction to create amylose. Note to comp: Set figure in this order: N (alpha-glucose image) → amylose image + $(n-1)H_2O$

chromatos/Shutterstock.com

---

[3] "Carbohydrates" (n.d.) *Chemistry for Biologists*. Web. http://www.rsc.org/Education/Teachers/Resources/cfb/carbohydrates.htm . Accessed May 28, 2015.

sucrose forms a *dimer* (a polymer with only two parts), but the condensation reaction can be performed again and again to join sugars into polymers as well.

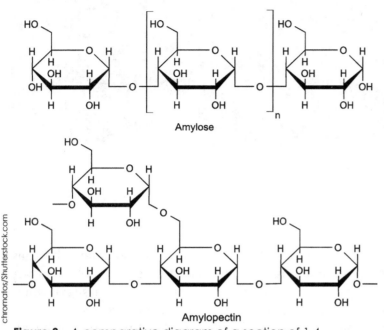

Amylose

Amylopectin

Figure 3 A comparative diagram of a section of 1,4 glycosidic-bonded amylose next to a section of 1,6 and 1,4 glycosidic-bonded amylopection.

chromatos/Shutterstock.com

Polymers of alpha-glucose—starches—usually are made in one of two main forms: in long chains that coil themselves neatly, known as *amylose*, pictured above; and in branched versions known as *amylopectin,* which are less space-saving but which are more efficient at providing energy more rapidly when the need arises. Amylose chains of alpha-glucose molecules are bound with alpha-glycosidic bonds, which dictate that the resulting polymers lie essential in linear chains. This arrangement allows for very efficient packing. The branched amylopectin also has glycosidic bonds, but between different neighboring carbons than those that bond in amylose.

Plants are generally understood to make both kinds of storage polysaccharides, in about a 1:4 ratio of amylose:amylopectin.[4] Digestion of polysaccharides in the human body requires the action of specific molecules called enzymes, which can break the glycosidic bonds. Enzymes are molecules that speed up the rate of a chemical reaction in a living organism.[5] In most cases, the enzyme molecules have names that are similar to the larger molecules that they affect. *Amylase* is the molecule that is responsible for reversing the condensation reaction that created amylose or amylopectin. Because the condensation reaction accomplishes polymerization through the removal of a water molecule between every two monomers, the reverse chemical reaction, *hydrolysis*, is recognizable by the addition of water into a polymer and results in the separation of the polymer back into its constituent monomers.

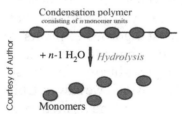

Condensation polymer
consisting of *n* monomer units

+ *n*-1 H_2O ↓ *Hydrolysis*

Monomers

Figure 4 An example of a hydrolysis reaction.

Courtesy of Author

The hydrolysis reaction does not happen all at once. It takes energy for the amylase enzyme to accomplish the chemical reaction. And certainly it takes **water**! Our need for dietary water is not just to keep our skin smooth, or to keep our nerves functioning well with dissolved ions, but also to make possible the hydrolysis of polysaccharides (and proteins), which provides us with the energy and materials necessary for life. The structure of the storage polysaccharide is a factor in the efficiency of amylase. Amylase is produced by the salivary glands in the mouth and by the pancreas for digestion in the intestines. Salivary amylase is not as productive as pancreatic amylase, but together they represent the ability of humans and many

[4] Bruso, Jessica (2015). "High Amylose Foods". *Livestrong.com*. Web. Accessed on May 28, 2015 at http://www.livestrong.com/article/465067-high-amylose-foods/

[5] "Definition of Enzyme" (n.d.) *MedicineNet.com*. Web. Accessed on May 28. 2015 at http://www.medicinenet.com/script/main/art.asp?articlekey=3266

other organisms to extract alpha-glucose from starches. This means energy for cells because the breakdown of alpha-glucose in the cells of the body results in energy for the body to function. The long chains of amylose are broken down much more slowly than the branched chains of amylopectin, and as such, some people find that blood sugar levels are more stable when they eat foods with higher amylose content, like corn, beans, some legumes (beans), and some varieties of rice.

What about in animals—what storage polysaccharides are produced by them—or rather, *us*? And why do we need storage polysaccharides, if we are not making sugar through photosynthesis? These are important questions. Animals end up with excess alpha-glucose from the breakdown of sucrose or other polysaccharide food sources like starch. Options for the storage of excess glucose exist for animals: store for short-term, rapid-access use as the polysaccharide *glycogen*, or shunt into the long-term storage, fat-creating process. At close range, molecules of glycogen look like amylopectin, with 1,4- and 1,6-glycosidic bonds. But pull away and see the larger picture and it is revealed that glycogen has twice as much branching than amylopectin. Another difference between glycogen and amylopectin is the size of the molecules themselves: glycogen molecules can consist of 60,000 alpha-glucose monomers[6] and amylopectin can be as large as between one and two *million* monomers (and amylose chains can be 500–20,000 monomers).[7]

What is the purpose of the higher degree of branching for animals than even the more branched amylopectin plant starch? Storage of sugars as polysaccharides is intended in all organisms for one purpose: energy access. The more branches in a storage polysaccharide, the more open chains are available for hydrolysis by molecules of amylase. And the more hydrolysis going on, the more sugars are being liberated, and the more energy is available for cells to use. The large muscle groups and brains (not to mention other complex energy-consuming systems) of animals require access to energy at a moment's notice: imagine a squirrel in the middle of the road, facing an oncoming car. The squirrel reacts quickly to avoid the car because its muscles have the ability to access glucose nearly instantaneously to fuel its brain and muscles. A clump of grass can't move out of the way of your foot as you move across the lawn because it lack the design and ability to do so, not the least of which is because it does not have a storage polysaccharide structure that can release its sugar fast enough (of course, it doesn't have muscles or a brain either, but that's beside the point).

So animal glycogen can be deconstructed by amylase enzymes at multiple points simultaneously. In animals, glycogen is concentrated in the liver and in skeletal muscles. Animal use of glycogen also benefits from the ability of glycogen to be hydrolyzed (this process is often called *glycolysis*) anaerobically, meaning it can occur even in the absence of available oxygen. This is important especially when animal muscles cannot get oxygen quickly enough to fuel normal

Figure 5

[6] "The Glycogen Molecule"(n.d.) Web. Accessed May 30, 2015. Available at http://www.biotopics.co.uk/JmolApplet/glycogen2.html

[7] Bowen m Richard (2006). "Dietary Polysaccharides: Structure and Digestion". Web. Access May 30, 2015. Available at http://www.vivo.colostate.edu/hbooks/pathphys/digestion/basics/polysac.html

respiration in cells.[8] In the liver, glycolysis occurs most often in response to demand for regulation of blood sugar levels.[9]

Structural Polysaccharides

Other types of polysaccharides also exist in the plant world. Polymers of *beta*-glucose are created for structural purposes in plants. Synthesis of structural polysaccharides is similar to that of storage polysaccharides in that the condensation reaction is performed as the method of polymerization, but the bonds formed between the glucose molecules in structural polysaccharides are *beta* linkages, which give the finished polymer a different spatial arrangement and orientation, and different physical properties than polymers of alpha-glucose with alpha linkages.

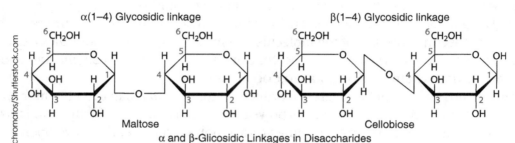

Figure 6   Difference between alpha and beta Glycosidic linkages.

Structural polysaccharides don't have the linear, tend-to-coil, chains like storage polysaccharides. Beta-linked polysaccharides have a more 'extended' structure,[10] which does not tend to coil but lends itself to more rigidity and strength. This explains why plants can stand up against gravity, reaching for the sun's rays. In many plants, these chains of beta-glucose polymer are identified by the name *cellulose*.

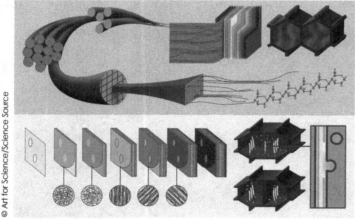

Figure 7   Cell wall structure.

This rigidity is enabled by the hydrogen bonding that happens along a polysaccharide chain as well as between chains. (Storage polysaccharides have some of this hydrogen bonding as well, though the alpha-linkages orient the sugar monomers differently and lessen the impact of hydrogen bonding as a rigidity factor.)

Cellulose is the major component in textiles like cotton, linen, jute, and any other plant-based textile (known also as *vegetable fibers*).[11] Cellulose textile materials offer many attractive

---

[8] "Glycogen Metabolism Notes" (2009). Web. Accessed on May 30, 2015. Available at http://oregonstate.edu/instruct/bb450/summer09/lecture/glycogennotes.html

[9] King, Michael W. (n.d.) "Glycogen Metabolism". Web. Accessed May 30, 2015. Available at http://themedical biochemistrypage.org/glycogen.php#catabolism

[10] Chaplin, Martin (2014) "Polysaccharide Structure" *Water Structure and Science*. Web. Accessed on May 30, 2015. Available at http://www1.lsbu.ac.uk/water/polysaccharides.html

[11] "Natural Cellulose Fibres" (n.d.) *Textile School*. Web. Available at http://www.textileschool.com/articles/70/natural-cellulose-fibres-natures-own-fibres

qualities: strength, absorbency, flexibility, softness, among others. Since ancient times, people have been cultivating cellulosic plants for use after harvesting in clothing and other household materials, building materials, and paper. Why are cellulosic fibers so versatile? Hydrogen bonding is a huge factor. Consider a cotton bath towel: soft and strong—hydrogen bonding between polysaccharide threads give the flexibility that brings softness and the molecular integrity that keeps the towel in one piece for years of use and washing. Towels are expected to be absorbent—hydrogen bonding defines the attraction of the exterior hydroxide groups on the sugar monomers and the water molecules on your skin after a shower. Towels are dyed many colors—the hydrogen bonding that exists in the woven fabric also lends itself to bonding to molecules that have color (more on this in Drugs and Dyes).

Lignin is a secondary component of plant cell walls, especially so in 'woody' plants. Lignin is also a polymer, though not a polysaccharide; it is an irregular aromatic alcohol polymer whose purposes seem to be both structural partnership with cellulose and also protection from microbial degradation.[12,13] As this diagram reveals, the main aromatic units are molecules whose structures look similar to aromatic spice compounds.

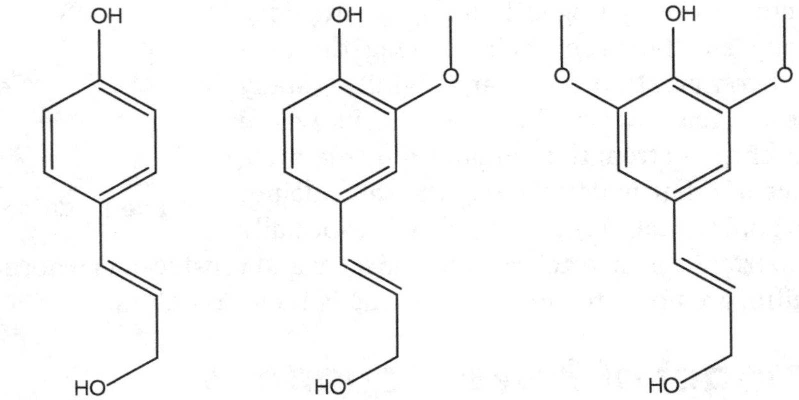

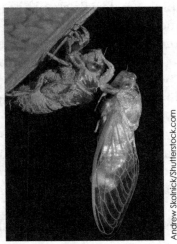

p-Coumaryl alcohol Coniferyl alcohol Sinapyl alcohol

Courtesy of Author

Figure 8 It should not be surprising that plants are able to build many different compounds from the same basic starting materials.

Other structural polysaccharides besides cotton exist, and one of the most important is a cellulosic derivative called chitin (pronounced *kite-in*). Not a plant-based material, chitin is exactly the opposite: it is created by creatures that have an exoskeleton, like crustaceans (like crabs, shrimp, etc.) and insects (spiders, ants, locusts, etc.).

The structure of chitin is nearly exactly like cellulose, but there is a change of functional group on the carbon #2 of the beta-glucose monomer:

Andrew Skolnick/Shutterstock.com

Figure 9 Locust shedding its exoskeleton.

chromatos/Shutterstock.com

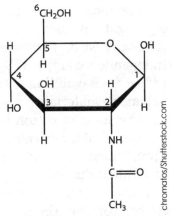

Figure 10 *N*-Acetyl glucosamine, chitin monomer.

---

12 Bartley L, Ronald P. 2009. Plant and microbial research seeks biofuel production from lignocellulose. Cal Ag 63(4):178–184. DOI: 10.3733/ca.v063n04p178. Available at http://californiaagriculture.ucanr.edu/landingpage.cfm?article=ca.v063n04p178&fulltext=yes#

13 Vanholme R[1], Demedts B, Morreel K, Ralph J, Boerjan W.(2010) "Lignin biosynthesis and structure". *Plant Physiology,* July 2010: 153(3): 895–905. Doi: 10.1104/pp.110.155119. Epub 2010 May 14. Accessed on May 30, 2015 from http://www.plantphysiol.org/content/153/3/895.full.pdf+html

This chemical composition change results in a structural polymer that is also rigid and strong like cellulosic materials (think trees), but the hydrogen bonding interactions within and between polymeric chains are different because of the nitrogens and carbonyl oxygens now present in the chains.[14] Different, how, you ask? Look at the structure of adjacent monomers after polymerization:

Notice that the monomers arrange themselves so that the large nitrogen-carbonyl groups (*N-acetyl groups*) are on opposite sides of the equator of the polymer chain. These larger side groups (compared to the – OH hydroxyl groups on regular beta-glucose) increase the hydrogen bonding interactions along and between chains, giving the exoskeleton great hardness and rigidity. (Many insects have composite exoskeletons where the chitin's strength is enhanced by the presence of other materials, like the salt calcium carbonate, $CaCO_3$.) This rigidity is especially

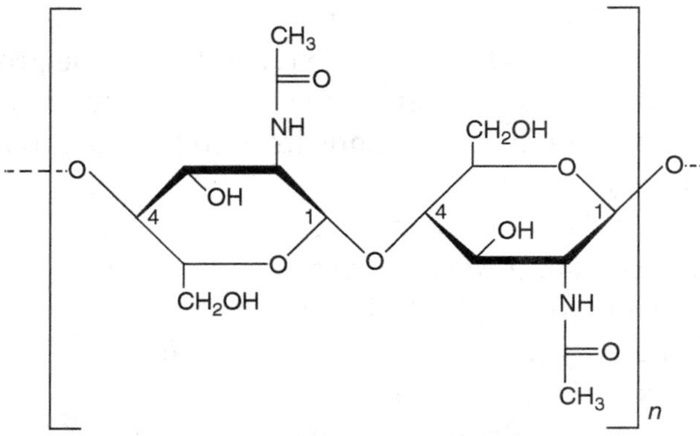

Figure 11 Image of section of chitin.

important for invertebrates like insects and crustaceans whose soft inner tissues, if not covered by chitin, are no match for the weapons of their predators.

Impact of Polysaccharides

It has been suggested that polysaccharides were the first natural polymers which formed on Earth.[15] Cellulose is not only the most prevalent polymer compound on earth, it is also used for many commercial products like clothing, paper, film, explosives, and plastics.[16] Though modern man-made polymers are replacing natural cellulosic materials in many modern products, the historical, cultural, and political importance of cellulosic materials should not be underestimated.

The most important historical American cellulosic polymer textile is cotton. America's discovery might have been inspired by visions of shortened, profitable trips to the spice-rich Orient, and America's earliest commercial product was sugarcane, but cotton fueled the America of the War of Independence (1774–1783) and the Civil War (1861–1865). The knowledge accumulated during the Industrial Revolution of the 19th century when 'Cotton [was] King'[17] prepared America to create new polymeric materials in the 20th century.

The value of cotton in the young America represented, except for sugarcane and tobacco, the most profitable agricultural crop. Though Columbus found cotton crops growing in the Bahamas in 1492,[18] the installation of European colonies in North America brought cotton to what would become the United States 300 years later. Initially, American cotton was illegal to import to Europe, as a protection for the powerful industry of English wool of the time; but by the start of the 16th century, cotton crops were being cultivated in the Virginia colony. (Spanish colonists had already been raising

---

[14] Kameda, T., Miyazawa, M., Ono, H. and Yoshida, M. (2005), Hydrogen Bonding Structure and Stability of α-Chitin Studied by ^{13}C Solid-State NMR. Macromol. Biosci., 5: 103–106. doi: 10.1002/mabi.200400142.

[15] V. Tolstoguzov, Why are polysaccharides necessary? *Food Hydrocolloids* **18** (2004): 873–877.

[16] "Cellulose" (2014). *Science Clarified*. Web. http://www.scienceclarified.com/Ca-Ch/Cellulose.html

[17] Gates, Henry Louis Jr. (2013). "Why Was Cotton 'King'". *100 Amazing Facts About the Negro*. Web. Available at http://www.pbs.org/wnet/african-americans-many-rivers-to-cross/history/why-was-cotton-king/

[18] "The History of Cotton" (n.d.) Web. Available at http://www.cotton.org/pubs/cottoncounts/story/index.cfm

cotton in Florida for 50 years!).[19] Initial crop yields were low, mostly due to inexperience among the settlers. (In comparison, regions of the Middle East, India, and South America had cultures where cotton production was fully established). As colonies became more established, cotton plantations grew as well, which necessitated more labor hands. The slave trade, already firmly entrenched in the sugar industry, also supplied the cotton industry with a 'free' labor force (as ironic as that sounds) to keep profits high for His Majesty the King. As the 17th century moved into the 18th, America pulled into independence, and changes in seed type and cotton processing technology positively impacted cotton production. Interestingly, though the idea of slavery was waning as the American Revolutionary period approached, the value of cotton as a domestic product for the new young country was viewed as too critical to let the slave labor force go. Prior to the American Revolution, Britain was the main market of American cotton, in part due to the climate of England: the dampness in the air reduced cotton fiber breakage during processing of the raw fibers into thread; and the concentrated skilled (and unskilled) labor force already present in the British Isles.[20] When the American Revolution ended, Britain had been importing nearly 23 million pounds of American cotton. The domestic American market for cotton had been growing, however, during the same period. Spinning cotton into thread and weaving cotton textiles for domestic military uniforms and other uses became a symbol of national pride. After the war ended, American cotton manufacturing began in earnest, which continued to depend on cotton from the American southeast, which was possible at those levels of production only because of the slave population.

British consumption of American cotton continued nearly unabated for the next 150 years. British finishing of raw cotton was still better than American factories could provide, at least in terms of speed. The invention of the cotton gin at the end of the 18th century further enabled consumption of cotton products at high human cost. The cotton gin (*gin* is an abbreviation of *engine*) is a mechanical separator, which more efficiently cleaves the raw cotton fibers from the rest of the field plant material. Until 1793, cleaning the picked cotton was a human task, and even the most efficient hand-cleaning could only produce about 1 pound of cleaned cotton lint per day. The American cotton yield at that time was no more than 156,000 bales of cotton annually, and about 700,000 slaves lived in the country. But the cotton gin (created by Eli Whitney as a response to the bottleneck he observed on the Georgia plantation where he worked as a tutor) changed *everything*.[21]

By the end of the first decade of the 19th century, cotton production and slave population had both grown dramatically. The earliest hand-cranked cotton gin improved cotton-cleaning yield 50-fold, and it wasn't long until larger gins appeared (powered by leather belts and gears turned by teams of oxen or mules, and then steam engines) and yields became larger and more financially profitable. More profits led to larger plantations and more slave labor to work them. As

Figure 12 Early example of a hand-cranked cotton gin.

© Bettmann/CORBIS

[19] "The Story of Cotton" *Cotton's Journey*. Web. Available at http://cottonsjourney.com/Storyofcotton/page2.asp

[20] "The Industrial Revolution" (n.d.) *History.com*. Web. Available at http://www.history.com/topics/industrial-revolution

[21] Dattel, Eugene R. (2006). Cotton, the Global Economy: Mississippi (1800–1860)". Available at http://mshistorynow.mdah.state.ms.us/articles/161/cotton-in-a-global-economy-mississippi-1800–1860

America crested the early lip of the Civil War, the production of cotton had grown to over 400 million pounds annually, and the slave population was nearly four million.[22] And, even though slave trading had become illegal in the United States in 1808, intracontinental slave trading was rampant, with free blacks kidnapped and transported across state lines into states where slavery continued (North and South Carolina, Mississippi, Florida, Alabama, Georgia, Louisiana, Texas, Virginia, Arkansas, and Tennessee[23]) for which financial success rested on the backs of slave laborers, legally 'obtained' or no. And so the American Civil War broke out over differences in the social and economic frameworks of the North and South over income sources (North: diversified industry; South: cotton agriculture), slavery (North: no; South: yes), stance on abolition of slavery (North: yes; South: no), and acceptance of the election of Abraham Lincoln to the presidency (North: yes; South: no).[24] When Abraham Lincoln declared the Emancipation Proclamation in 1863, immediately freeing all slaves in states that were in rebellion (the Confederate States), the position of the southern labor force became even more tenuous, and the outcome of the war was all but assured to favor the Union. Couple this with skilled sea blockades in the Atlantic Ocean aimed at preventing supply imports to Confederate states and at preventing southern cotton from reaching British factories for final processing,[25] and the war ended with Union victory in 1865.

Everett Historical/Shutterstock.com

Figure 13

The American Civil War occurred within the larger historical period known as the Industrial Revolution. Historians place the start of the Industrial Revolution in Britain at just about the time that America gains her independence from King George III.[26,27] As the name suggests, this period in history is a great time of innovation and growth due to improvements in technology, especially in the textile and iron industries. Of particular note is the improvements to the steam engine (originally developed in 1712), attributed to the work of James Watt, and this machine went on to power the rest of the Industrial Revolution. And while it is true that the living conditions for many improved over the course of the Industrial Revolution, the poor and working classes did not benefit hardly at all, at least at first. Factory jobs were better paying than starving on a country farm, but the urban living conditions were squalid at best, and disease was rampant. Employers felt no compulsion to provide for their workers anything but a job, and workers came in all ages. Since schooling was not made compulsory by the state, anyone who could be taught a factory skill and could work 12–14 hours each day[28] to bring money into the household. In British cotton refineries, for example, children as young as five were employed to untangle threads inside the toothy gullets of looms because their hands

[22] "Total Slave Population in United States, by State" (n.d.) Web. Available at http://thomaslegion.net/totalsla veslaverypopulationinunitedstates17901860bystate.html

[23] "Confederate States of America" (2014). *Wikipedia*. Web. Available at http://en.wikipedia.org/wiki/ Confederate_States_of_America

[24] Kelly, Martin. (n.d.) "Top Five Causes of the Civil War" Web. Available at http://americanhistory.about. com/od/civilwarmenu/a/cause_civil_war.htm

[25] Surdam, David G. (1998). "The Union Navy's blockade reconsidered". *Naval War College Review* **51** (4): 85–107.

[26] "King George III & The American Revolution" (n.d.) Web. Available at http://totallyhistory.com/ king-george-iii-the-american-revolution/

[27] "The Industrial Revolution" (n.d.) *History.com*. Web. Available at http://www.history.com/topics/industrial-revolution

[28] "Childhood Lost: Child Labor during the Industrial Revolution"(n.d.). Web. Available at http://www.eiu. edu/eiutps/childhood.php

were small enough to fit between the spinning machinery. If a child or adult was injured, maimed, or even killed, that person was simply replaced and nothing more was done about it by the factory owner. People were dispensable.

The unethical treatment of workers did not fall completely on deaf ears. The Industrial Revolutionary period in America started a little later in America, but in the late years of the 19th century and early decades of the 20th century, the introduction of laws requiring a minimum working age, restrictions on the kind of work children could do, and the maximum number of hours a child could work per day improved the overall ethical treatment of children in America. Couple this with enactments of laws requiring compulsory minimum skills education for children between the ages of 7 and 16,[29] and the child labor force was definitely reduced. The seeds of labor unions to act as a collective voice for workers were planted with the organization of the National Labor Union in 1866, though it crumbled quickly; the Knights of Labor arose to take up the mantle in 1869. This latter group fought for an 8-hour workday and opposed child labor. The most long-lasting and successful labor organization was formed in 1886: the American Federation of Labor; joining forces in 1995 with the Congress of Industrial Organizations (CIO), together it become the AFL-CIO. Labor unions have diversified into organizations of specialty-skilled workers (e.g., the United Steelworkers (USW), or the International Brotherhood of Electrical Workers (IBEW)), but all are committed to representing their members to companies and the government insofar as it pertains to worker compensation, safety, and other benefits. It all began with the Industrial Revolution.

Rights for workers also encouraged women to voice a desire for representation in government through the right to vote. The right of black men to vote, ratified in the 15 Amendment in 1869, the right of women to vote, achieved in 1920 with the ratification of the 19th Amendment to the Constitution, and later the 1965 Voting Rights Act (which actually made possible the provisions of the 15th Amendment) are all connected to the Industrial Revolution and therefore to polysaccharides like cotton.

---

[29] "Compulsory Education" (n.d.) *USLegal.com*. Web. Available at http://education.uslegal.com/compulsory-education/

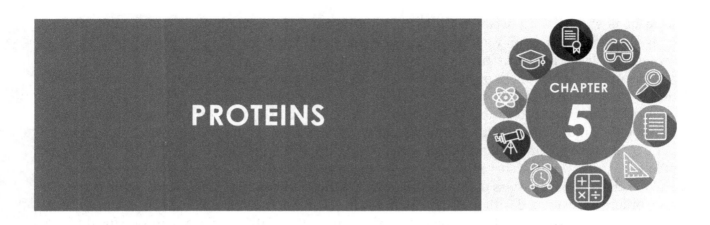

PROTEINS

CHAPTER
5

This unit started out with a romantic description of spider silk. Spider silk, and that kind of fiber made by spiders, moths, and butterflies primarily, is actually a polymer that is most accurately classified as a *protein*. Other materials, like muscle tissue, are also proteins. All proteins are used for structural purposes, in transport of materials, as enzymes, or to open and close cells as channel proteins. And all proteins on earth have something in common: In order to be a protein, a material is made of polymers of *amino acids*. What is an amino acid? I am glad you asked.

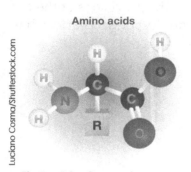

Amino acids

Luciano Cosma/Shutterstock.com

Figure 14 General structure of an amino acid.

Amino acids are compounds that contain nitrogen—the *amino* prefix is a nod to that—and also an organic acid functional group—a carboxyl group (C=O) with a hydroxyl group on it. These parts of an amino acid molecule flank a central carbon, which has at least one hydrogen on it as well:

The 'R' group represents any general organic side group on the central carbon: anything from a single hydrogen, to more complex carbon groups. Because carbon binds so well to other carbons, there are nearly infinite possibilities for 'R', but nature has produced about 500 different variations here on earth[30] (or at least that is as many as science has identified so far). Of those, only 20 of them are used in biological life here on earth to make all proteins on earth.

Proteins are created from the instructions encoded in a cell's genetic material, DNA. The DNA is like a cookbook that has a recipe—also called a *gene*—which describes the protein we want to make. The DNA cookbook is written in a special language of nucleotide *bases*—there are only four of these: guanine, thymine, cytosine, and adenine. All the words in the DNA cookbook are composed of 'letters', which are groups of three nucleotides. Each trio of nucleotides is a code word, which represents an ingredient of the protein: an amino acid.[31] The first phase of protein synthesis, called *transcription*, involves copying the DNA recipe from the genetic cookbook located in the nucleus of a cell. Then the copied recipe is brought to the part of the cell outside the nucleus where proteins are made, the *ribosome*—this is kind of like the kitchen of the cell. All proteins are made—this second

---

[30] Wagner I, Musso H (November 1983). "New Naturally Occurring Amino Acids". *Angew. Chem. Int. Ed. Engl.* **22** (22): 816–828. doi:10.1002/anie.198308161

[31] "Protein Synthesis" (2014) Web. Available at http://www.bbc.co.uk/schools/gcsebitesize/science/add_edexcel/cells/synthesisrev1.shtml

Basic Amino Acids

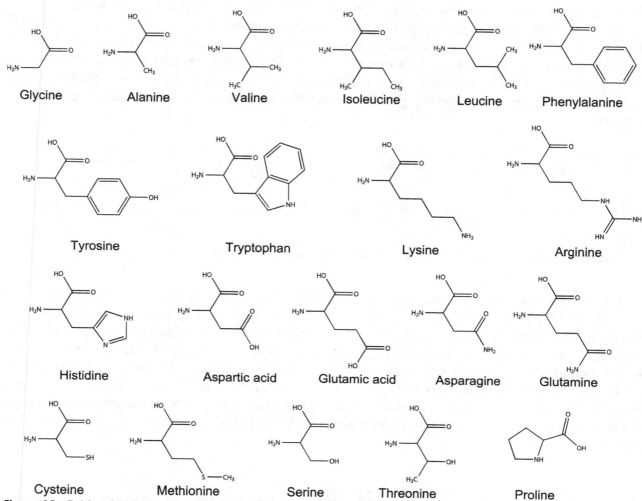

Glycine Alanine Valine Isoleucine Leucine Phenylalanine

Tyrosine Tryptophan Lysine Arginine

Histidine Aspartic acid Glutamic acid Asparagine Glutamine

Cysteine Methionine Serine Threonine Proline

Figure 15 Table of 20 amino acids used in living creatures on earth.

chromatos/Shutterstock.com

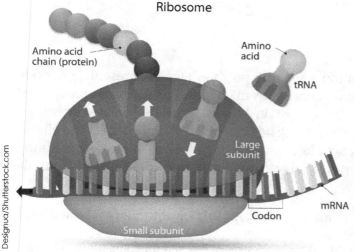

Ribosome

Amino acid chain (protein)

Amino acid

tRNA

Large subunit

Codon

mRNA

Small subunit

Designua/Shutterstock.com

Figure 16 Schematic of protein synthesis.

Ermolaev Alexander/Shutterstock.com

phase of protein synthesis is called *translation*—with instructions copied from DNA in the nucleus but made in the ribosome kitchen of the cell using amino acids and chemical reactions.

All proteins are polymerized with the same chemical reaction: the condensation reaction! This time, the reaction brings together amino acids end to end in a specific orientation—acid end of one amino acid aligned to the amino end of another amino acid—so that water is produced as a side product and the bond that is created for the protein polymer is known as a *peptide* bond. All peptide bonds are composed of a carboxyl group bound to an amino group and are only found in proteins:

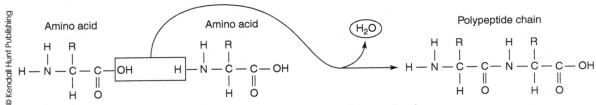

Figure 17   Condensation reaction accomplishing a peptide synthesis.

Proteins can vary in length, depending on the DNA recipe instructions, where the protein will be used and how it will be used. In every case, the finished protein will have a three-dimensional configuration, called its *conformation*, which is crucial to the proper operation of the protein in its particular role in the system. For example, insulin, the molecule responsible for the ability of cells to take in and use glucose for energy, two short polymer chains tethered together. In mammals, one chain, the *A chain*, is composed of 21 amino acids and the other, the *B chain*, has 30 amino acids.[32] (Interestingly, when insulin is in storage in the pancreas, groups of six polymers are held in one bundle by a zinc ion—a zinc-polymer *complex*; when being used, the insulin is a single A–B pair.[33])

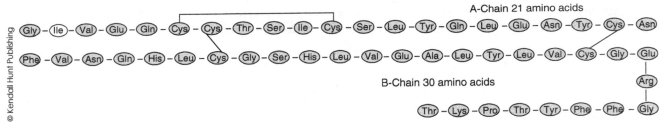

Figure 18   Amino acid sequences of insulin polymer.

Each short connection between a pair of amino acids in a polymer chain represents a peptide bond formed via one incidence of a condensation reaction. How many molecules of water are formed during the production of the two chains of insulin? (20 in chain A and 29 in chain B). The larger lines connecting three pairs of cysteine amino acids (Cys) are *disulfide bonds*, which connect the chains and bring integrity to the overall pair, making it act like a single unit. (Disulfide bonds are common in proteins; in fact, hair (a protein called *keratin*) which is curly is just that way because of disulfide

---

[32] Cartailler, Jean-Philippe (n.d.) "The Structure of Insulin". Web. Available at http://www.betacell.org/content/articleview/article_id/8/page/1/glossary/0/

[33] Chang X, Jorgensen AM, Bardrum P, Led JJ (1997). "Solution structures of the R6 human insulin hexamer,". *Biochemistry* **36** (31): 9409–9422. doi:10.1021/bi9631069

bonds that hold the protein chain in coils. Straight hair has less disulfide bonding.) Of course, the conformation of insulin is not two-dimensional, as the listing of the amino acids on this page appears. In fact, it has a coiled, intertwining of the two chains:

So, we have established that proteins are polymers of amino acids, polymerized by the condensation reaction, and proteins are the compounds responsible for the majority of the most important functions and structures that make life possible on earth. Let's connect protein polymers with the textiles of the first section in this unit. What natural protein polymers are used in textiles (fabrics)? Silk and wool are the most common examples.[34] Silk is the fabric woven from strands of protein excreted by the larvae of the domesticated silkworm moth (species: *Bombyx mori*). What is interesting about this particular protein is that the majority of the protein strand is composed of a repeating pattern of six amino acids: glycine–serine–glycine–alanine–glycine–alanine–. This repeating segment is also known as *fibroin* and makes up 80–85% of all *B. mori* silk samples.

This repeating pattern of amino acids in silk gives the molecular structure that gives rise to the physical and chemical characteristics associated with silk: dye ability, softness, resistance to fiber distortion after stretching, visible sheen, lightweight, burn resistance, good insulating properties, ability to absorb moisture, just to name a

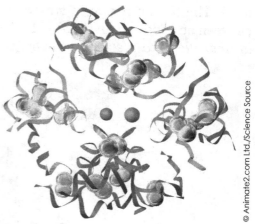

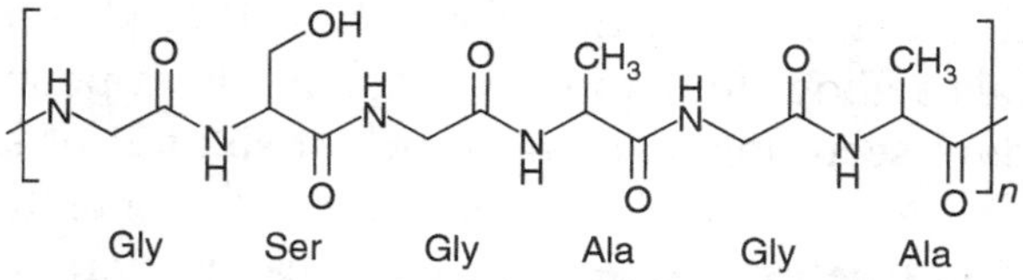

Figure 19 3-D Structure of insulin.

Gly   Ser   Gly   Ala   Gly   Ala
Figure 20 Protein section of fibroin.

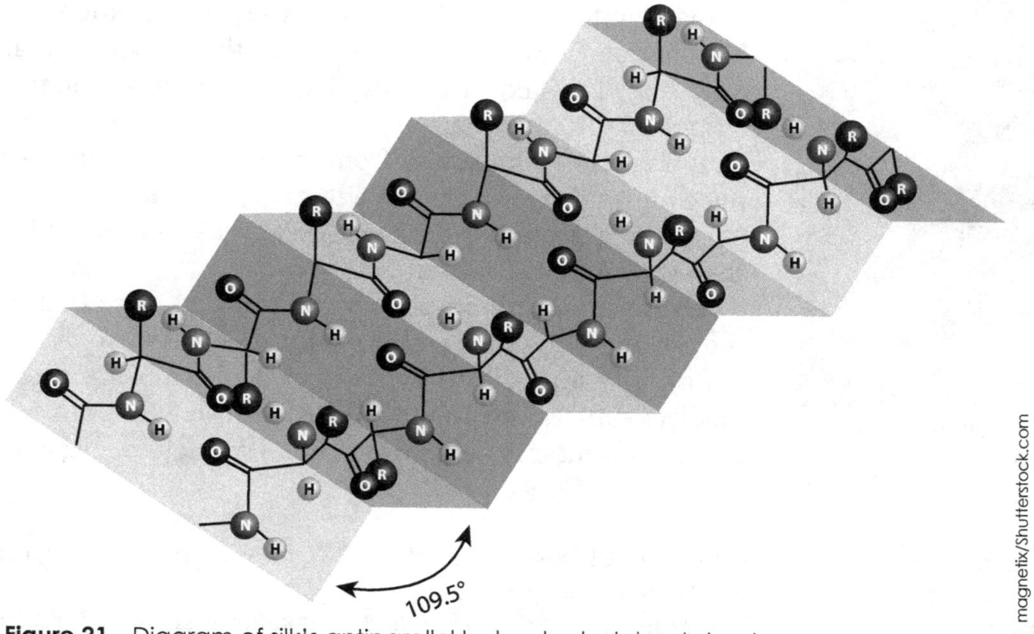

109.5°

Figure 21 Diagram of silk's antiparallel beta-pleated sheet structure.

[34] "Natural Protein Fibers" (2015) Web. Available at http://www.textileschool.com/articles/77/natural-protein-fibers

few.[35] The fibroin region in strands of silk protein encourages a great deal of hydrogen bonding between strands and gives the silk fibers, which can be as long as 5,000 amino acids in length, an *antiparallel beta-pleated sheet* structure:

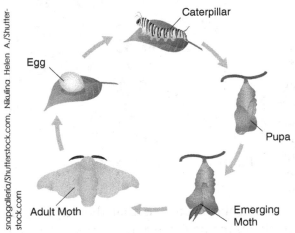

Figure 22   Life cycle of *B. mori*.

So, silk is amazing. But how is it made and harvested? *B. mori* silkworms, having been cultivated by humans for silk production for more than 5000 years, are completely domesticated; they are no longer found in nature and are completely dependent on humans for survival. The adult moths have evolved to have mouthparts that are degenerate—they cannot eat. They can't fly either—their wings are now too small to lift their fat bodies. The only drive of the adults is to reproduce.[36] They will die in a short time after emerging from the cocoon. The overall life cycle of the *B. mori* does not appear different than most moths and butterflies.

The female *B. mori* silkworm moth, once fertilized, can lay about 500 eggs. Hatched eggs release white caterpillars that eat *only* the leaves of the mulberry bush, and the caterpillars (called *silkworms* though they are not worms at all) will go through five growth stages where they shed their outer skin in order to grow larger. This process takes about four to six weeks, at which point the caterpillar stops eating and spends about three days secreting a sticky strand of silk out of special glands near its mouth, which it winds around itself to create a completely sealed cocoon. At this point, in the wild, caterpillars normally *pupate*, or go through the metamorphosis of changing from caterpillar to moth or butterfly. In the silk industry, called *sericulture*, most cocoons, once spun, are gathered and plunged into boiling water to 1)stop the pupation process and kill the pupa and 2) loosen the silk in the cocoon so that it can be unwound. The silkworm caterpillar weaves its cocoon from a single strand of silk, anywhere from 300 to 900 meters in length. This strand is simply unwound from the cocoon around the now-dead pupa and is rewound around a larger spool. Multiple strands (4–8) of cocoon silk are twisted together to make silk thread, which can then be woven into fabric.[37]

Commercial applications of silk include more than just fabric, though until the 20th century, silk was known only as a textile. Today, silk and silk protein fragments are being used to create medical sutures, man-made blood vessels, and more. Biologically, silk is nonimmunogenic (does not stimulate a negative immune

Figure 23   Harvesting of silk from cocoons.

[35] Ballenberger, Walt (2007) "Properties and Characteristics of Silk" Web. Available at http://ezinearticles.com/?Properties-and-Characteristics-of-Silk&id=488797

[36] Damodaram, K. J. P., Kempraj, V., Aurade, R. M., Rajasekhar, S. B., Venkataramanappa, R. K., Nandagopal, B., and Verghese, A. (2014). Centuries of domestication has not impaired oviposition site-selection function in the silkmoth, *Bombyx mori. Scientific Reports, 4,* 7472. doi:10.1038/srep07472

[37] "Silk Making and Silk Production" (n.d।). Web. Available at https://texeresilk.com/article/silk_making_how_to_make_silk

system response), nontoxic, and has broad medical applications to many animal species,[38] including humans. Silk beauty treatments are also popular, with lotions, shampoos, and cosmetics being marketed with silk (or said to be 'silk-like') to improve softness, shine, or return a youthful appearance to the skin or hair.

Silk is applicable today, but is a product of antiquity. Sericulture's long history began in China, where cultivation of silkworms is recorded in written accounts more than 5000 years old. The amount of labor required in sericulture makes silk an economic investment—it takes more than 3,000 silkworms to create a kilogram of silk—and this translates to a higher market price for silk than for some other textiles, like cotton—and the value of the fabric is reflected by those that have worn it: royalty and aristocracy, the wealthy and the powerful.[39] China had a monopoly on silk and silk production in the ancient Far East, and demand for this product by other nations and peoples gave rise to a network of trade routes spreading through the Far East and Middle East, which are

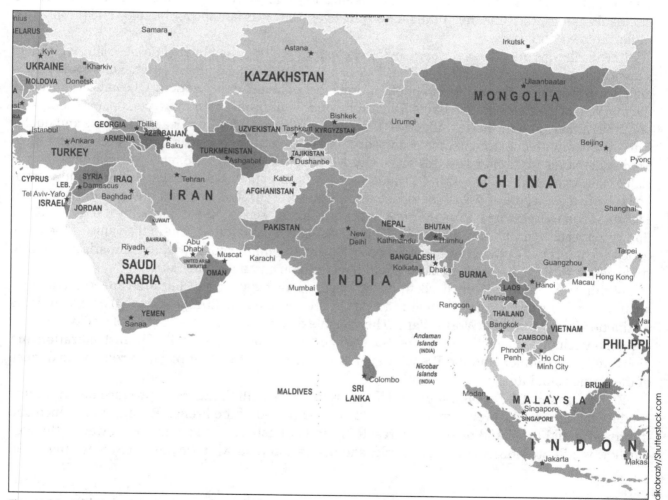

Figure 24   The Silk Road map.

[38] Hao, Zhang, Li Ling-ling, Dai Fang-yin, Zhang Hao-hao, Ni Bing, Zhou Wei, Yang Xia, and Wu Yu-zhang. 2012. "Preparation and characterization of silk fibroin as a biomaterial with potential for drug delivery." *Journal Of Translational Medicine* 10, no. 1: 117–125. *Academic Search Complete*, EBSCO*host* (accessed June 1, 2015).

[39] Edlund, Erica (n.d) "Silk- Thin as a Human Hair, Stronger than Steel". Available at http://www.kth.se/che/kemi2011/2.27954/matfeb-1.79866

collectively known as the *Silk Road*.[40] It should be noted that the Silk Road trade routes were responsible for more than silk commerce, but also of many other products we have already encountered: solid sugar from Arabia and Indonesia, spices from India and the Spice Islands, silk from China, and more. How do you think these products ended up in Venice for soldiers and merchants to haggle over during the Middle Ages? The Silk Road, which was a set of overland routes from Asia to Europe, and its commercial sea- and water-based trade route companion the Spice Road (often they were the same roads) brought traders and products back and forth for thousands of years.

The desire for silk brought strangers to China's door. China's dynastic government realized the value of the silk product and of sericulture and made export of live eggs or caterpillars a crime punishable by death.[41] Nonetheless, the traffic along the Silk Road included opportunists who were willing to take the risk. Starting around 200 BCE, sericulture spread outside China's borders, first to Korea, then to Turkey and India, and by 200 AD, Japan had its own sericulture industry as well. Europe joined the sericulture industry just about 200 years before the Arab invasion of the Iberian Peninsula, and by the time the Middle Ages were over, Italy was famous for its silks, and it remained so for the next 500 years.

Italian silk was popular in Europe—being cheaper (in theory) than buying Chinese silk, which had logistical costs covering the entirety of the Silk Road and then some. Italian silk merchants commanded high prices from the aristocratic and noble patrons that demanded the product. Of course, Venice was the marketplace for all silk merchants from everywhere. But profit can make a person, or a company, or a government, greedy. Italian silk prices were going up steadily for European aristocrats and nobles who were committed to dressing themselves only in silk, in order to underscore their societal standing. French demands for lighter and less expensive silk created an untenable business relationship with Italian silk merchants; as a result, in 1466, King Louis XI established the domestic silk industry in Tours, with a main goal of trying not to just produce the best silk, but to wean France off of Italian silk. The addiction was costing France 400,000–500,000 golden ecus a year at this point.[42] If the assumption is that each ecu was approximately 3.5 grams of gold at about that time,[43] the value of the trade deficit with Italy has a modern value of between $15 and $19 *million* today.[44] (For comparison, the U.S. trade deficit with China for all goods as about $350 million annually, and rising.)[45]

Within 100 years, the French silk industry was thriving, having been chartered anew as a monopoly industry in Lyons by King Frances 1 in 1535. France became the European capital of silk, and this distinction lasted until the World War I. The variable demand for silk, the impact of a silkworm disease that decimated French sericulture, the changing labor force as a result of industrialization and global military conflict, and the French Revolution were all factors that played a role in the demise of the French silk industry.

But it's not as if we should mourn. The pressure on the silk industry to produce an affordable product for the masses grew exponentially during the period of the French Revolution. French silk improved in lightness and quality as a result of market push. At the same time, however, the cost of silk was still prohibitive for most people, and disillusionment with the monarchy was intensified

[40] Whitfield, Susan. 2015. *Life Along the Silk Road*. Oakland, California: University of California Press, 2015. *eBook Collection (EBSCOhost)*, EBSCOhost (accessed June 1, 2015).

[41] Fei Xiu Qin (2006). "The Spread of Sericulture". Available at http://www.ancientsites.com/aw/Article/827025&about=ContentFeatures&aboutData=827025

[42] Georges Duby (ed), *Histoire de la France: Dynasties et révolutions, de 1348 à 1852* (vol. 2), Larousse, 1999 p. 53 ISBN 2-03-505047-2

[43] "15th Century" (n.d.) Web. http://info.goldavenue.com/Info_site/in_arts/in_mill/15thcentury.htm

[44] "Gold Price" (n.d) Web. http://goldprice.org/gold-price-per-gram.html

[45] "Trade in Goods with China" (2015) Web. http://www.census.gov/foreign-trade/balance/c5700.html

due to changes in philosophy (coming out of the Enlightenment) about what made a good government as well as more immediate problems with getting enough food to eat. These factors, among others, led to the overthrow of the constitutional monarchy and an establishment of the First French State in 1792.[46] Napoleon's *coup d'etat* in 1799 brought a different kind of authoritarian regime to France, which coincided with the birth pangs of the Industrial Revolutionary period. By the time a monarchy returned to France in 1815, there were irreversible changes in philosophy that affected everything about French society, industry, culture, and *life*. The Napoleonic Code had planted the ideas of *liberty, equality, and fraternity* for all French citizens.[47] Most notably, the poor were not content to stay under the thumb of the rich and powerful. Throughout the 19th century (and into the 20th century), French government underwent the throes of its own labor as the country reinvented itself. The rise of the common people into positions of power as landowners and business owners changed the way business was done and the investment in the technology that was booming around them all over Europe.

The textile industry had its own metamorphosis. The silk weaving industry experienced upheaval when the programmable loom came into existence and made the human weaving assistants unnecessary. This same technology is implicated in the development of computers within a century.[48] Pressure for quality silk at nonsilk prices also pushed the innovation of artificial silk-like fabrics like rayon, nylon, polyester, and more. Today's synthetic fiber industry has its roots in the French Revolution. Additionally, a bacterial epidemic decimated the French silkworm population around the same time that America was emerging from the Civil War. Silk supply dwindled, and there was an economic impetus for a replacement.

Rayon, a remanufactured natural cellulose polymer, was first to arrive, as the world approached the 20th century, as the Industrial Revolution's momentum was waning. Initially created from sawdust waste from lumbermills, rayon is a fiber created from the repolymerization of beta-glucose polymer shreds. Rayon starts with a thick—*viscous*—solution of wood pulp and water. This is very inexpensive starting material! Chemical treatment of the slurry ends with *extrusion* through nozzles, which creates strands of cellulose pulp, and then these strands are passed through an acid bath, which enables the condensation reaction to recreate longer chains of cellulose. The properties of cellulose are still in force (light, soft, absorbent, etc.), but the processing of remanufactured strands can give the fibers a more lustrous, silk-like appearance. 'Viscose rayon' products were a good start to silk replacement textiles, but there are disadvantages: viscose is not as strong as its natural cellulose or natural silk counterparts, especially when wet.[49] Rayon burns easily, shrinks easily (and permanently), and is not impervious to chemicals. But boy is it affordable over silk! Today, raw silk can be purchased as cheaply as $2–$15 per yard; finished silk can be more than 20 times that price![50,51] Rayon can be 1/10 the price, even for the most intricately woven patterns.[52] The appearance of rayon made 'silk' affordable to more people, and inspired efforts to create more artificial fibers.

---

[46] Schwartz, Robert M. (n.d.) "History 151: The French Revolution: Causes, Outcomes, Conflicting Interpretations". Available at https://www.mtholyoke.edu/courses/rschwart/hist151s03/french_rev_causes_consequences.htm

[47] Alexander Grab, *Napoleon and the Transformation of Europe*(2003), as referenced in Wikipedia "History of France" (2015), available at http://en.wikipedia.org/wiki/History_of_France#cite_note-80

[48] Essinger, James (2004) *Jacquard's Web*. Oxford University Press.

[49] "Viscose Rayon" (n.d.) *Swicofil.com*. Available at http://www.swicofil.com/viscose.html

[50] *Online Fabric Store* Available at https://www.onlinefabricstore.net/content-silk.aspx#sort=salepricedesc

[51] *Alibaba.com* Available at http://www.alibaba.com/showroom/raw-silk-fabrics-price.html

[52] *Fabric.com*. Available at https://www.fabric.com/apparel-fashion-fabric-rayon-fabric.aspx?sort=Price+%28Descending%29

As technology and chemical knowledge increases collided, more ideas for new textiles emerged. In the first few decades of the 20th century, interest in artificial polymers became a research interest and then an industry of its own. Companies like DuPont created whole divisions of *research and development (R&D),* aimed solely at discovering and creating novel materials and compounds, rather than manufacturing and producing finished products. This concept, of a company to invest money in R&D, which has a mixed potential for profit, was novel in and of itself, but has now become the standard in all kinds of business. Collecting bright minds together and engaging them in pursuits which may seem open-ended has led to some very profitable ends: textiles like nylon, Kevlar, and polyester; plastics like polypropylene, polyethylene, and polyvinyl chloride (PVC); and products like lasers, satellites, and cell phones, just to name a few.

Nylon's production was one of the first products birthed from R&D at DuPont. Wallace Carothers, venerated and immortalized in bronze at the DuPont installation in Wilmington, Delaware, created the first nylon compounds in 1934 (he is also the creator of neoprene rubber, in 1931). His background was in accounting, but World War I opened up a teaching position in chemistry at Tarkio College when the regular instructor was drafted. He fell in love with the field, and after the war ended, Carothers went to graduate school, earning a Ph.D. in chemistry in 1924, studying the infant field of polymer chemistry. DuPont hired him in 1928, and the rest is history. Carothers' brilliance in chemistry was in contrast to his deep struggle with depression; in 1937, he committed suicide after losing his sister suddenly, leaving behind a young wife and daughter. The world can only speculate what other wonders he would have created, given the opportunity.

Nylon is a wonderful condensation polymer, created after observing the chemical structure of proteins in textiles like silk. Carothers reasoned that, if amino acids will condense to form polymeric proteins that other peptide-like bonds should be able to be created the same way. However, nylon differs from silk's protein structure in that the monomers are larger than two carbons, and, rather than having an amino end and an organic acid end on each monomer, the monomers would be flanked by either two acid groups or two amino groups. During polymerization, the monomers would arrange themselves in the most favorable way in order to effect the condensation reaction between monomers with differing end groups. The end result is a stable polymer consisting of short chains of carbons connected to one another by *amide bonds.* Amide bonds are the same as peptide bonds: a carbonyl group (−C=O) connected to an amino group (−NH); but the term *peptide* is reserved for use only in proteins—compounds of amino acids. Carothers' nylon benefits from the stability of the amide bond, the tendency of the polymer chains to fold in beta-sheets, and the creation of hydrogen bonds along and between chains. You can see the similarity to silk in these characteristics. The processing of nylon also lends itself to smooth, shiny fibers, making it a great artificial silk. And

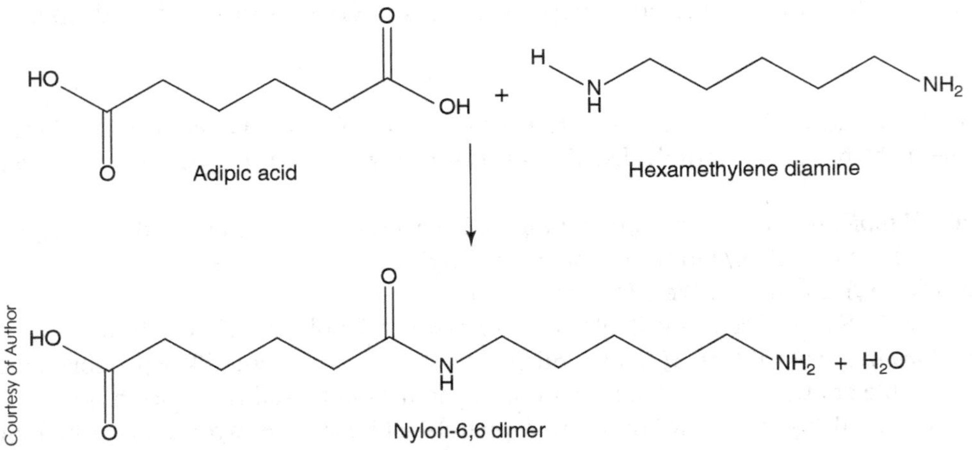

Figure 25 Condensation of the monomers of nylon 6,6.

because there are fewer amide bonds in a polymer chain of nylon than there are peptide bonds in a silk segment of the same length, there are about half as many hydrogen bonds in a sample of nylon. This might sound like a disadvantage, but nylon is still a very strong fiber. The advantage of fewer hydrogen bonds is a lesser tendency to absorb and hold water, which expands the applications of nylon over silk.

The world of polymers exploded out of the work of chemists like Carothers. DuPont's legacy of R&D has also led to the creation of Kevlar and Nomex, created by the condensation reaction between different monomers than those of nylon. Kevlar and Nomex, created through the work of DuPont's chemist Stephanie Kwolek, are synthetic aromatic polyamide polymers, also known as *aramids*. These types of polymers have the commercial application in bulletproof vests and in protective equipment used by firefighters. The rigidity of the fibers in Kevlar provides dispersion of close-range impact forces, and the high melting point of Nomex is due to the high molecular stability of the crystalline

Figure 26   Kevlar and Nomex polymer structures.

structure in the fibers themselves. See the similarity in the amide bonds of aramids to those in nylon. The aromatic rings between the amide bonds also bring an increased stability to the overall aramids, greater than that of the carbon chains of nylon compounds.

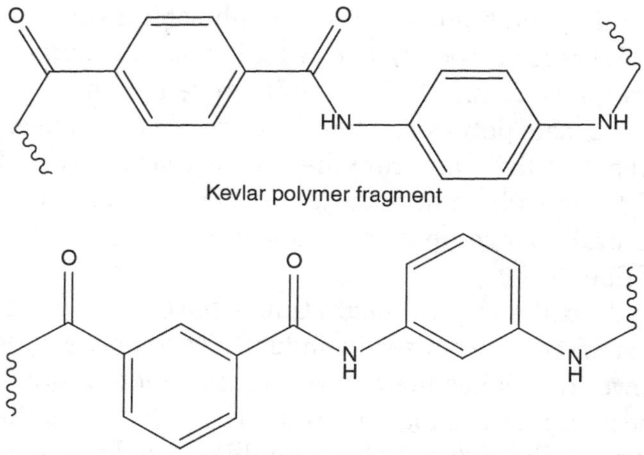

Finally, this discussion of artificial fiber textiles would not be complete without including *polyester*. The term implies something about the chemical structure of the polymer. Rather than an amide linkage, these kinds of polymers have *ester* bonds between the monomers. An ester bond has the general formula of

$$R - \overset{\overset{\displaystyle O}{\|}}{C} - O - R'$$

where R and R' represent two organic carbon groups. Esters are often created with the condensation reaction occurring between an organic acid functional group on one monomer and an organic alcohol group, a hydroxyl group, on the next monomer. This is most easily facilitated when each monomer has an organic acid group on one end and a hydroxyl group on the other. Sound a little familiar? This is a little like the amino acid structure and polymerization in proteins. Alternatively, monomers can have two organic acid groups on either end, or two hydroxyl groups on either end, reminiscent of nylon's two acids groups with two amino groups. The structures of R and R' can be varied, lending different qualities to the final finished products. For example, in thinking about polyester fabrics, R and R' are usually aromatic rings and short carbon chains:

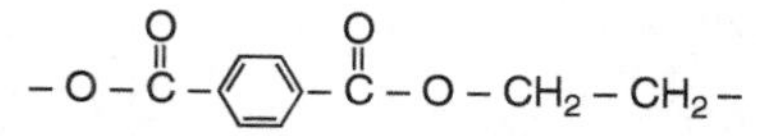

Figure 27   Polyethylene terephthalate.

This polymer is actually *polyethylene terephthalate*. The monomers are terephthalic acid $HO - \overset{\overset{\displaystyle O}{\|}}{C} - \langle \rangle - \overset{\overset{\displaystyle O}{\|}}{C} - OH$ and ethylene glycol HO-CH$_2$-CH$_2$-OH (remember this molecule from the discussion about sweet molecules in Unit 1?). Can you see how the molecules of water are formed from each pair of monomers? If a polyester chain was formed from the condensation of 5,000 monomers of terephthalic acid with 5,000 monomers of ethylene glycol, how many ester bonds are created? The commercial value of polyester cannot be underestimated. Polyesters of more types than just polyethylene terephthalate (PET) are on the market, and these materials are used in textiles for clothing and home fabrics like upholstery, window treatments, bed linens, and stuffing for pillows; in fibers for carpeting and rugs; in ropes, conveyor belts, safety belts, and much, much more. Polyester, having very little to no hydrogen bonding sites, offers some advantages over nylons or other condensate polymers: stain and wrinkle resistance, durability and high color retention. These physical properties stem from the chemical structure and lend themselves well to combination with those of natural fibers to form textile blends, which maximize the advantages of both the artificial and natural fiber components while minimizing the disadvantages of a textile made of only a single type of fiber.[53]

In some ways, though, things have come full circle for silk. The Chinese silk monopoly, which was first compromised by industrial espionage 2300 years ago, has returned. Today, the 21st century power of silk lies mainly with China, and with India as a distant second. These two countries have the manpower to together produce nearly 98% of the world's silk. (Interestingly, though China is the largest silk producer, about 1 million people are employed in the sericulture industry there; in India, nearly 7.9 million people produce a silk output only about 20% of that of China. Perhaps the technological prowess of China is part of the impressive efficiency of the sericulture industry there.)[54] China is also a world producer of artificial textiles of many types, so leaking the secret of silk 2000 years ago may have been the best thing that ever happened for China.

---

[53] *Wikipedia.org*. "Polyester" (2015). Available at https://en.wikipedia.org/?title=Polyester
[54] "Statistics" (2013). Available at http://inserco.org/en/statistics

RUBBER

No discussion of polymers would be complete without including *rubber*. Rubber is so common in modern society that it is hardly given a second thought. Rubber bands and rubber pencil erasers in school and the office, rubber gaskets and rubber tires on the car, rubber balls and rubber soles on sneakers in the sports arena, rubber gloves in the medical field, and rubber boots in the rain, and rubber threads in the elastic in our underwear. The properties that give rubber such immense commercial application and value are what they are because rubber is also a polymer. Not a protein, or a polysaccharide, but instead a different kind of polymer whose monomer is different than any we have seen yet: *isoprene*.

$$H_2C = \overset{\overset{\displaystyle H}{|}}{C} - \underset{\underset{\displaystyle CH_3}{|}}{C} = CH_2$$

This two-dimensional structure is an incomplete picture of the molecular structure, but nonetheless, it is obvious that this monomer is not like any others we have seen so far: no oxygens, no organic acid groups, and no hydroxyl groups. Therefore, the conclusion that a polymer made from this monomer would not be formed via condensation would be correct. Polymers that are formed from the isoprene monomer are created through *addition polymerization*. This type of polymerization does not require the loss of any atoms or molecules in order to connect the monomers; instead, the bonds within a monomer will rearrange in order to allow electrons to form new bonds with neighboring monomers. In some materials, this process occurs spontaneously. This is true for natural-sourced rubber and for styrene, the starting material for the polymer used in Styrofoam. For other materials, many of them man-made, this process must be enabled by the presence of other materials like catalysts. This would be the case for materials like polyethylene (PE), polyvinyl chloride (PVC), polypropylene (PP), acrylics, polyvinyl acetates, and for commercial products like Teflon and Saran wrap.

TABLE OF ADDITION POLYMERS

Monomer	Polymer	Uses/applications
Isoprene $H_2C = C - C = CH_2$ with H above second C and CH_3 below	Poly-isoprene	Tires, elastics, reinforcing materials, consumer clothing and home goods, automotive parts, etc.!
Styrene $HC = CH_2$ (attached to benzene ring)	Polystyrene $-[-HC-CH_2-]_n-$ (attached to benzene ring)	Plastic cutlery, DVD cases, plastic combs and toys, foam packaging materials[55] thermoset
Ethylene (also known as ethane) $H_2C = CH_2$	Polyethylene $-[-HC-CH_2-]_n-$	Thermoplastic plastic drinking bottles, formed plastic tanks and pipes, low friction machinery, plastic shopping bags[56]
Vinyl chloride (also known as ethyl chloride) $H_2C = C$ with H and Cl	Polyvinyl chloride $-[-H_2C-C-]_n-$ with H above and Cl below	Credit card material, plastic plumbing, toys, flooring, rigid plastic bottles, medical equipment and devices[57]
Propylene (also known as propene) $H_2C = C$ with H and CH_3	Polypropylene $-[-H_2C-C-]_n-$ with H above and CH_3 below	Thermoplastic plastic parts for the automotive industry, toys, plastic laboratory tools, carpeting and upholstery, specialty paper, sound equipment[58]
Acrylonitrile $H_2C = C$ with H and $C \equiv N$	Polyacrylonitrile $-[-H_2C-C-]_n-$ with H above and $C \equiv N$ below	Fibers for textiles like clothing and carpeting. Also used as co-polymer in synthesis of carbon fiber[59,60]
Tetrafluoroethylene (also known as tetrafluoroethene) $C = C$ with F, F, F, F	Polytetrafluoroethylene (also known as Teflon) $-[-C-C-]_n-$ with F, F above and F, F below	Nonstick coatings for cookware and cooking utensils[61]

[55] "What are the different uses of polystyrene?" http://www.wisegeek.com/what-are-the-different-uses-of-polystyrene.htm

[56] Sondalini, Mike. "Polyethylene, it's properties and uses" *Feed Forward Publications* http://www.slideshare.net/bin95/183-polyethylene-itspropertiesanduses

[57] "How is PVC Used?" http://www.pvc.org/en/p/how-is-pvc-used

[58] Johnson, Todd. (n.d.) "What is Polypropylene?" http://composite.about.com/od/Plastics/a/What-Is-Polypropylene.htm

[59] "Polyacrylonitrile" (2003) http://pslc.ws/macrog/kidsmac/polyac.htm

[60] "Addition Polymers" (n.d) http://elmhcx9.elmhurst.edu/~chm/vchembook/401addpolymers.html

[61] "Addition Polymers" (n.d) http://elmhcx9.elmhurst.edu/~chm/vchembook/401addpolymers.html

Vinylidene chloride (also known as 1,1-dichloroethylene) $H_2C = C \begin{smallmatrix} Cl \\ Cl \end{smallmatrix}$	Polyvinylidene chloride $-[-H_2C-\underset{Cl}{\overset{Cl}{C}}-]_{n}^{-}$	Plastic food films like Saran wrap, seat covers, material in some clothing or imitation leather goods[62,63]
Vinylacetate $H_2C = C \begin{smallmatrix} H \\ O-C-CH_3 \\ \parallel \\ O \end{smallmatrix}$	Polyvinyl acetate $-[-H_2C-\underset{O-C-CH_3}{\overset{H}{C}}-]_{n}^{-}$ $\overset{\parallel}{O}$	Adhesives (glues), paper coatings, paints[64,65]
Vinyl alcohol $H_2C = \underset{OH}{CH}$	Polyvinyl alcohol $-[-H_2C-\underset{OH}{CH}-]_{n}^{-}$	Textile sizings, adhesives, paper coatings, strengthening agent for cement products[66]

Structures Courtesy of Author

What should be obvious about all the monomers in the Table of Addition Polymers is that each of them has a short chain of two carbons double-bonded together. And, in the process of polymerization, this little chain loses one of the bonds in the double bond, but gains bonds to monomers on either side. Clearly, there is something about this double bond, which lends itself well to polymerization. Each chemical bond, regardless of where it is found in whatever compound it is found in, is composed of two electrons. These pairs of electrons are shared in organic compounds between the atoms on either end of the bond. When pairs of electrons bundle in groups of two (double bond) or three (triple bond), the connection between the terminal atoms is even stronger. But because electrons are the source of all the chemistry that happens (except for nuclear chemistry), having more than one pair of electrons between atoms means that there is some reactivity in the situation as well. In the case of polymers like these, the double bonds can be broken and electrons will move to bond with new atoms. This causes a cascade effect, enabling more and more monomers to bond to the polymer. Initiation of the cascade may have a number of sources. As mentioned previously, some compounds like styrene and isoprene can self-polymerize, meaning that environmental factors like sunlight or other natural sources can initiate the polymerization cascade. In most other addition polymers, initiation is enabled by catalysts or by the presence of chemicals like peroxides.

[62] "Addition Polymers" (n.d) http://elmhcx9.elmhurst.edu/~chm/vchembook/401addpolymers.html
[63] Reusch, William (2013) "Polymers: https://www2.chemistry.msu.edu/faculty/reusch/VirtTxtJml/polymers.htm
[64] "Poly(vinylacetate)" (2005) http://pslc.ws/macrog/pva.htm
[65] "Vinyl Acetate Monomer (VAM)" (n.d.) http://www.lyondellbasell.com/Products/ByCategory/basic-chemicals/Acetyls/VinylAcetateMonomer%28VAM%29/index.htm
[66] Marten, F. L. 2002. Vinyl Alcohol Polymers. Kirk-Othmer Encyclopedia of Chemical Technology. Available at http://onlinelibrary.wiley.com/doi/10.1002/0471238961.2209142513011820.a01.pub2/abstract

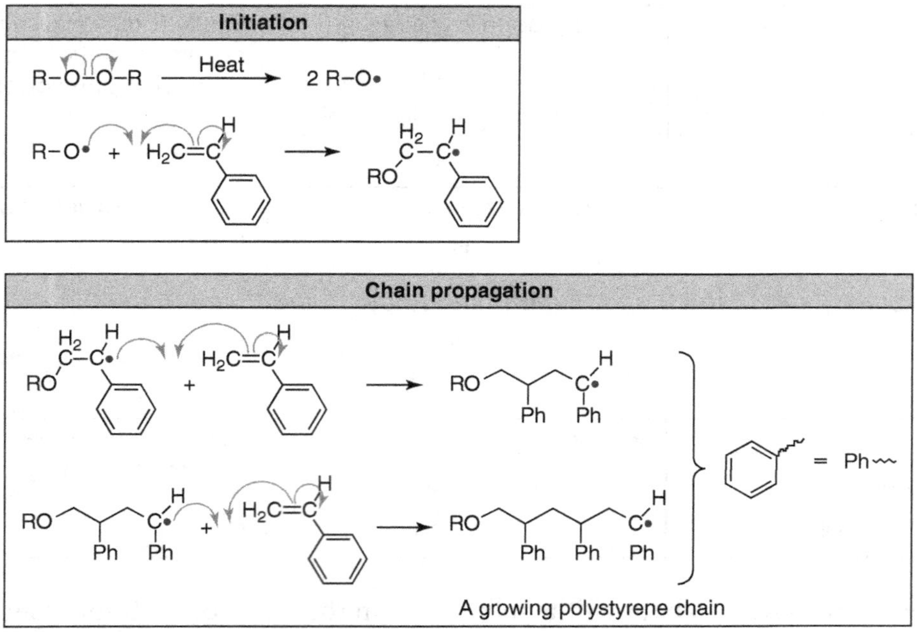

Initiation

$$R\text{-}O\text{-}O\text{-}R \xrightarrow{\text{Heat}} 2\ R\text{-}O\bullet$$

Chain propagation

A growing polystyrene chain

$= Ph\sim$

chromatos/Shutterstock.com

The importance of materials like rubber cannot be underestimated. What is interesting is that, despite the simplicity of the molecular structure presented so far, isoprene is more than it appears. The double bonds, which are the source of all the chemical power in the molecule, actually dictate the 3-D structure of the entire molecule. Because of this, there are actually *two* different isomers of isoprene: *cis*-isoprene, which has the double bonds oriented in the same direction along the center axis of the compound:

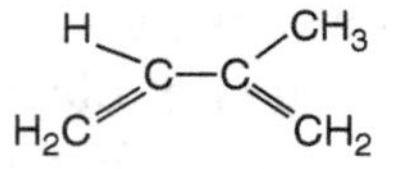

And *trans*-isoprene, which has the doubly bonded pairs on opposite sides of the central axis:

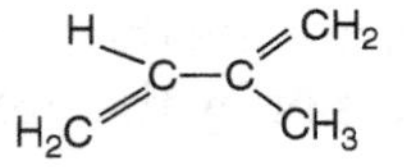

Polymers of isoprene vary depending on the isomeric orientation of the monomer used. Poly-*cis*-isoprene is a flexible, elastic polymer, *an elastomer*—the name should imply the stretchiness most people associate with rubber bands: the ability to stretch several times its original length and then spring back to that original length and shape. Poly-*cis*-isoprene chains can do this because the polymer arranges the monomers in an alternating fashion, which lends the overall molecule a structure that takes up quite a bit of space:

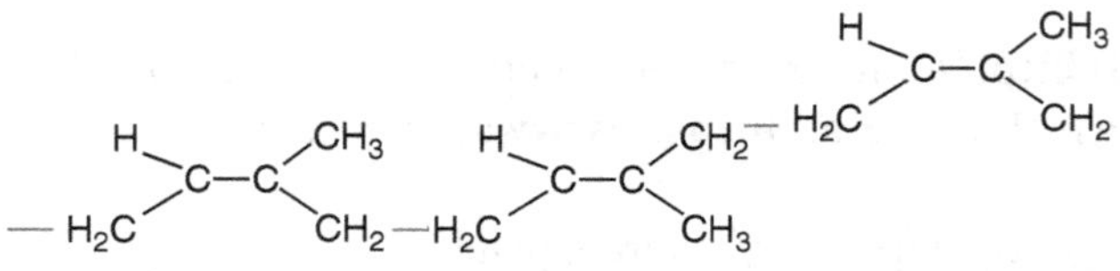

Poly-*cis*-isoprene

Courtesy of Author

The need to alternate the location of the $-CH_3$ group on each monomer by rotating the monomers 180° for each addition results in chains of polymer that do not pack tightly together in the final product. The means there is room to slip past one another, and that the weak forces which hold them together (not hydrogen bonds, as there are no −OH or −NH groups) are easily overcome. The next time you stretch a rubber band to its breaking point, this will remind you of the chemistry behind it.

In contrast, polymers of isoprene using the *trans*-isoprene monomer are hard, tough, and leathery, but also longer-lasting under stress than their *cis*-isoprene counterparts. The monomer arrangement is different than the *cis*-based polymer, which gives rise to that difference in physical properties. Trans-isoprene polymer chains do not have to rotate in order to polymerize. This means that each polymer chain exists in a relatively compact sleeve in space, and that many chains of polymer can arrange themselves in a pretty tight spatial arrangement. Even those weak attractive forces which exist are compounded by the physical nearness of the chains. This gives poly-*trans*-isoprene the toughness and rigidity, which are expected of it.

Courtesy of Author

Which, of course, means that different poly-isoprene products are used for different purposes. Trans-isoprene-based polymers are better for things like rubber stoppers and gaskets, which must be more impervious to chemical attack or physical degradation.

Rubber's convenient physical properties have been known for hundreds of years. Certain species of plants contain poly-isoprene in the sap. Brasil's rubber tree, *Hevea brasiliensis*, has been tapped for its sticky sap, beginning with ancient Olmec, Mayan, and Aztec peoples who would make rubber balls and rubber shoes by evaporating and shaping the sticky sap over smoky fires.[67] The Age of Discovery brought the material to Europe, and the rest is history, as they say. The Industrial Revolutionary period owes its success in part to rubber. Innovation in engineering of machinery and construction was enabled by rubber gaskets and hoses, and tires, not to mention rubber pencil erasers and rubberized boots and raincoats and umbrellas. Styrene is also a plant-originating material, created by the *Liquidamber* trees within the *Hamamelidaceae* family.[68] The word styrene is a reference to the words *styrax balsam*, a name given to the resin (sticky, wax-like sap) from trees of this family. This substance has a sweet smell and has been used as an incense ingredient, as a wine flavor additive, and as a hair perfume by the ancient Greeks.[69]

Today's poly-isoprene rubber and polystyrene are no longer solely sourced from plants. Monomers of these materials are obtained often from the specific chemical treatment of raw petroleum through the process of *refining* and *cracking*. Petroleum mining, regardless of its political implications and economic impacts, is a main source of starting materials for polymers like poly-isoprene rubber and polystyrene, as well as pharmaceutical, beauty products, and petroleum fuel products, and much more. Refining of petroleum is the name given for the process of taking raw materials from petroleum refining (crude oil, natural gas, shale oil, etc.) and separating these mixtures along boiling point

---

[67] "Rubber Tree: hevea brasiliensis" *Rainforest Alliance.* http://www.rainforest-alliance.org/kids/species-profiles/rubber-tree
[68] "Styrene" *Wikipedia.org. (2015).* https://en.wikipedia.org/wiki/Styrene
[69] "Styrax balsam" *Wikipedia.org.* (2015). https://en.wikipedia.org/wiki/Styrax_balsam

lines into materials like kerosene, gasoline, diesel fuels, propane gas, lubricant oils, waxes, petroleum jelly, and more. Using boiling point to separate a mixture of substances is called *distillation*.

Cracking is the process of breaking large organic compounds from petroleum sources into small organic compounds. Given the world's consumption of Styrofoam as an example—55 billion pounds worldwide in 2010—reliance on only plant sources for styrene is unsustainable and irrational. Cracking raw petroleum materials into styrene, or isoprene, increases the world supply more than plant-based sources alone could ever provide.

So polymers are all around us and within us. Life as we know it would not exist without polymers; actually, life would not exist at all without polymers. Polymers are many and varied. The inspiration for many of our modern man-made polymers came from nature, and the polymers we make also impact nature the other way as well. The environmental impact of polymers is as significant as the polymers themselves. In recent decades, for example, the proliferation of plastic grocery bags makes a trip to the market easier, but these bags also fill up landfills and do not decompose readily. The low production cost and ease of use of plastic grocery bags have landed them in more than 80% of the world's grocery and convenience stores, leading to a consumption of between 500 million and a trillion plastic grocery bags each year.[70] Polyethylene grocery bags, while having a relatively simple chemical structure, will still require, under optimum conditions, a plastic shopping bag, will need 500–1000 years to decompose.[71]

In the 1980s, the transition to plastic bags was evident in the question posed by the supermarket checkout clerk: 'Paper or plastic?' These three little words have had a different impact on the world than other three-little-word phrases like 'I love you'. We might have been better off, considering our landfills, with the latter group.

---

[70] Roach, John (2003) "Are plastic grocery bags sacking the environment?" http://news.nationalgeographic.com/news/2003/09/0902_030902_plasticbags.html
[71] Sleight, Kenneth (2011) "How Quickly does Plastic Degrade?" http://www.brighthub.com/environment/green-living/articles/107380.aspx

DRUGS AND DISEASE

Ring around the rosie,

A pocket full of posies

Ashes, ashes,

We all fall down!

- Children's nursery rhyme, first published in 1881

The nursery rhyme 'Ring around the Rosie', which has several variations, has as its inspiration the bubonic plague, also known as the Black Death. This disease appears to have been spread to Europe from India in the 1330s,[1] killing as much as 60% of the region's population; estimates put the worldwide death toll at 75 million.[2] This wasn't the first time that this disease consumed the majority of Europe and Asia's population: there was a recorded pandemic during the 6th century AD, which is said to have killed 'half of Europe'.[3] Legend has it that the disease originated first in China and was initially spread with Mongolian traders trading silk and other goods along the Silk Road and Spice Road.[4] And this disease has surfaced sporadically since then, with recent tests indicating that the bacterium responsible for the disease, *Yersinia pestis,* is the ancestor of all bubonic plague outbreaks over the centuries it has been known to be active.[5] The Black Plague did

[1] McGovern, John F., and John Alan Ross. "Black Death in Europe." *Salem Press Encyclopedia* (January 2015): *Research Starters*, EBSCO*host* (accessed July 4, 2015).

[2] Winchester, Simon. "'The Black Death', John Hatcher's Remarkable History of the Plague." (June 25, 2008). Book Review. *The New York Sun*. Available at http://www.nysun.com/arts/the-black-death-john-hatchers-remarkable-history/80591/.

[3] McNeil, Donald G. "No Longer Leading Killer, Plague Still Raises Fears." *The New York Times* (January 27, 2014). Available at http://www.nytimes.com/2014/01/28/science/no-longer-leading-killer-plague-still-raises-fears.html?_r=0.

[4] Berry, Gail. "'Ring Around A Rosie' A Brief History of the Bubonic Plague." (2013 Orca Health). Available at http://healthdecide.orcahealth.com/2012/08/21/ring-around-a-rosie/.

[5] Brownstein, Joe. "Bacteria causing 'Black Death' likely Extinct, Study Finds." (August 29, 2011). *Livescience.com*. Available at http://www.livescience.com/15826-black-death-bacteria-extinct.html.

not arrive in the United States until 1900, with the last known epidemic in the states occurring in 1924–1925. Since that time, the plague rarely occurs, and when it does, it is most likely to do so in the American southwest (New Mexico, Colorado, and Arizona) and along the Pacific coast states of California and Oregon.[6]

What is Disease?

The Black Death is so named because of the blackening of the skin where the *buboes* or swollen lumps appear. The disease has several possible courses of progression, from hours to days to weeks between the onset of symptoms and the end of the illness either through death or recovery. At its most basic definition, *disease* is a condition where a body is impaired from its usual function, either in a specific region or as a whole (often through the manifestation of pain or weakness), and is generally not caused by a direct physical injury but instead is caused by environmental factors like malnutrition, climate, or chemical agents; by infectious agents like worms, bacteria, or viruses; by genetic factors; or by any combination of these factors.[7,8]

Most people think of disease as something that can be treated with medicine, which can either remove the cause of the disease (as in a drug which kills a bacteria responsible for the symptoms of the disease) or ease the symptoms while a person or other organism allows the disease to 'run its course' (e.g., taking a cold medicine to ease the symptoms of fever or runny nose). Medicines are considered beneficial when they bring healing. Medicines in the United States are generally classified as *drugs*, and most drugs are produced by pharmaceutical companies that must adhere to strict rules for quality set by the Food and Drug Administration (FDA). The FDA has been operating in one form or another since 1906 and is charged with 'protecting and promoting public health through the regulation and supervision of food safety, tobacco products, dietary supplements, prescription and over-the-counter (OTC) pharmaceutical drugs (medications), vaccines, biopharmaceuticals, blood transfusions, medical devices, electromagnetic radiation emitting devices (ERED), cosmetics, animal foods and feed, and veterinary products.'[9] This is a huge responsibility and carries with it huge ramifications. Generally speaking, there is an assumption that the FDA is doing its job, and therefore the medicines we buy at the pharmacy or the grocery store will have been thoroughly investigated for their benefits and will result in improvement in overall health for a person taking those drugs or using those products.

What Kinds of Helpful Drugs Are There?

Most people understand that drugs are chemicals. Some are helpful, some are harmful. Some of the most commonly used drugs today have their roots in traditional medicines used by indigenous people of a certain geographic area, and some drugs are completely new. One thing that all substances we call 'drugs' have in common is that they all have an impact on the body; or in other words, the body has a response to the presence (and sometimes the absence) of these chemicals.

---

[6] "Plague in the United States" (n.d.) *Centers for Disease Control and Prevention*. Available at http://www.cdc.gov/plague/maps/.

[7] "disease" (n.d.) *Merriam-Webster online dictionary*. Available at http://www.merriam-webster.com/medical/disease.

[8] "disease" (n.d.) *Biologiy-online.org*. Available at http://www.biology-online.org/dictionary/Disease.

[9] "Food and Drug Administration" (July 2, 2015). Available at https://en.wikipedia.org/wiki/Food_and_Drug_Administration.

The decision to use a particular drug often arises out of a human need for a solution to a bodily, physical problem, while some may be used as an emotional escape from problems or to enhance a spiritual experience. For example, as humans have been on the earth, they have searched for ways to alleviate pain. For example, many cultures have chewed the bark of the willow tree (*Salix* genus) in order to extract the compound *salicylic acid*, which acts as an *analgesic*, relieving pain (Figure 1).[10]

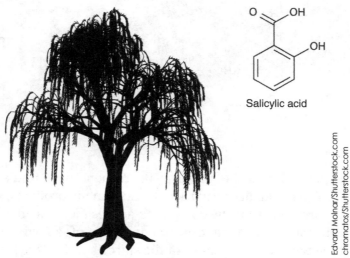

Salicylic acid

Edvard Molnar/Shutterstock.com
chromatos/Shutterstock.com

Figure 1   Image of weeping willow tree, inset of molecular structure and name of salicylic acid molecule.

There are many other examples of natural sources providing substances that have a biological impact, be it beneficial or harmful. For example, the opium poppy creates the alkaloid molecules morphine and codeine, the leaves of *Eucalyptus* provide an important cough syrup ingredient, and the bark of the cinchona tree provides an antimalarial medication. And plants aren't the only sources of important bioactive compounds; some animals are, too.[11] Many reptiles and amphibians create toxic skin secretions that act as nerve poisons. Some bacteria create compounds that paralyze nerves or cause massive gastrointestinal upset. In almost every case, these compounds are created by the plant or creature as part of a protective chemical arsenal aimed at preserving the chance of the organism to reproduce. Today's cultural interest in holistic medicines have also brought new awareness to some old remedies.

---

[10] Delaney, Terrence P. "Salicylic Acid" (n.d.), p 681. Retrieved from http://dept.ca.uky.edu/PLS440/PLS622/ **Salicylic acid**.pdf on March 25, 2015.

[11] "Sources of Drugs" (n.d.). Retrieved from http://howmed.net/pharmacology/sources-of-drugs/ on March 25, 2015.

ANALGESICS AND ANTISEPTICS

CHAPTER 7

Everyone has experienced pain, and most of the time, the response is to find a way to relieve it and to do it as quickly as possible. Today's analgesic market includes many products, not the least of which is *aspirin*. Aspirin is the chemical name for the compound acetylsalicylic acid, and it is a derivative – a compound that is chemically modeled after another compound – of salicylic acid, originally identified in plants in the family of the willow tree or those of the spirea species (Figure 2).

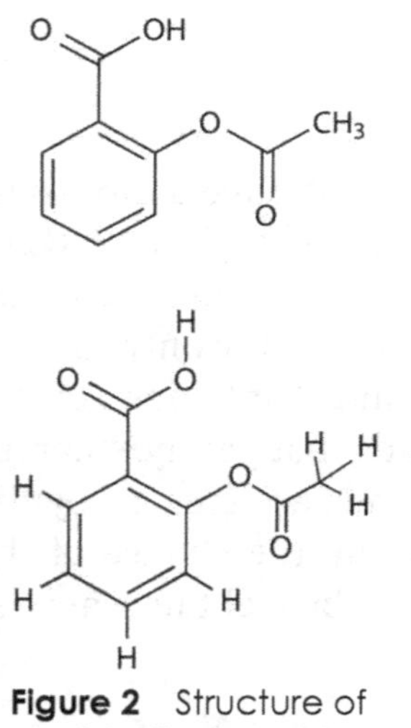

Aspirin was one of the first pharmaceutical analgesics, marketed initially by Bayer Chemical Company in 1900. Since then, aspirin has been found to have a number of other medical benefits, such as acting to reduce fever (being *antipyretic*), to reduce inflammation (an *anti-inflammatory*), and to reduce the ability of the blood to clot (helpful in some heart disease treatment plans). This last effect can also be a detriment if bleeding occurs from some other cause because presence of aspirin can extend the period of bleeding to dangerous levels. In addition, aspirin is widely known to irritate the stomach, causing minor symptoms like heartburn or, if taken too much or too often, serious problems like bleeding ulcers in the stomach. Aspirin has also been linked in some studies to complications when suffering from the chicken pox virus; this situation results in swelling in the brain coupled with liver malfunction, identified as *Reye* syndrome, which can be fatal. So do the benefits of aspirin outweigh the detriments? Yes. Aspirin's ability to soothe dull, distracting pain like a toothache, manage the peaks of a fever, reduce the swelling and discomfort of a sprained ankle, or thin the blood during a heart attack has saved more lives than it has cost.[12]

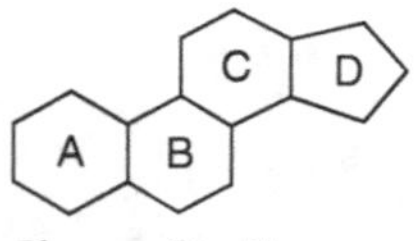

Some analgesic compounds are classified as *NSAIDS* – nonsteroidal, anti-inflammatory drugs. Aspirin falls into this category. So does ibuprofen. The label *NSAID* is a statement about the structure of the compound itself. Steroids have a particular structural backbone, containing four rings in a specific arrangement (Figure 3).

chromatos/Shutterstock.com

Figure 2 Structure of acetylsalicylic acid.

Courtesy of Author

Figure 3 The ABCD rings of all steroids.

A   B   C   D

---

[12] Liska, Ken (2000). <u>Drugs and the Human Body,</u> 6th Ed. Pp. 377–387 "OTC Nonnarcotic Pain Relievers: Aspirin, Acetaminophen, Ibuprofen, Naproxen Sodium, Ketoprofen, and Fenoprofen." Prentice Hall: Upper Saddle River, NJ.

It is obvious that aspirin does not have this structural backbone. Ibuprofen has this structure (Figure 4).

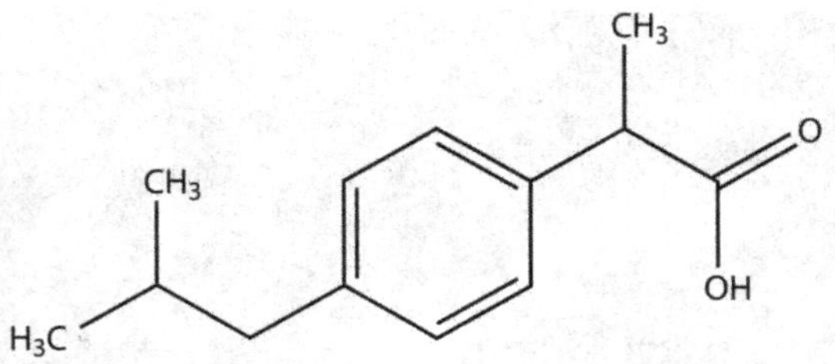

Both ibuprofen and acetylsalicylic acid are *phenolic* compounds: compounds that contain the functional group *phenol*. Phenol is an aromatic ring that has a hydroxyl group on it (Figure 5).

Many phenolic compounds exist, naturally and man-made. Phenolic compounds are often biologically active, causing a response in living systems. Many of the spices in the first unit are phenolic, for example.

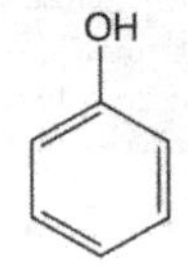

Phenolic compounds also have other known uses. Carbolic acid, an extract of coal tar, is actually a mixture of components but contains phenol as its main chemical component. Carbolic acid has *antiseptic* (germ-killing) properties, which were capitalized on by physicians like Joseph Lister during the 19th century who were trying to reduce the mortality rate. In some places, mortality after surgery in a hospital could be as high as 70%.[13] New ideas about the nature of disease being caused by invisible, living *germs* – an idea formulated by Louis Pasteur and supported by Lister, among others – was creating revolution in hospitals and homes around cleanliness. Substances like carbolic acid – really phenol – were changing how surgery was being done and how wounds were being cared for. Today's antiseptic molecules build on this history of phenols. Consider this phenols used in the popular mouthwash *Listerine*[14] (named for Joseph Lister!) (Figure 6).

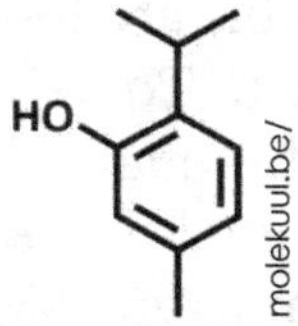

Just to be fair, though, some other molecules that have antiseptic action aren't phenols. Here are two, Figures 7 and 8.

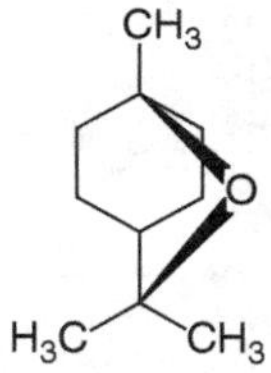

Figure 4   Molecular structure of ibuprofen.

Figure 5   Molecular structure of phenol.

Figure 6   Molecular structure of thymol.

Figure 7   Molecular structure of eucalyptol.

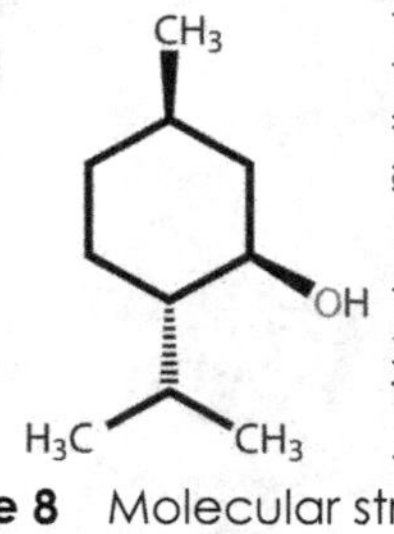

Figure 8   Molecular structure of menthol.

[13] LeCouteur, Jenny and Burreson, Jay (2003). Napoleon's Buttons. P. 114. Jeremy P. Tarcher/Penguin: New York.

[14] Thomas, Pat (2009). "Behind the label: Listerine teeth and gum defence [sic]." Available at http://www.theecologist.org/green_green_living/behind_the_label/269558/behind_the_label_listerine_teeth_and_gum_defence.html.

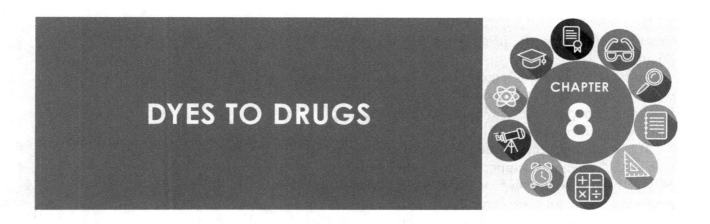

DYES TO DRUGS

CHAPTER 8

Phenols are found in other places as well. Compounds that are visually colored often are phenolic as well. To be perceived visually, three factors must be present: a physical object, a light source to illuminate the object, and an observer to see the object. Visible light in our solar system is provided naturally by the sun, and that visible light is just a small fraction of all the radiation in our universe. Visible light radiation is a type of energy, bundles of massless vibration that can be detected by our eyes' biological structures. Certainly, there is other energy in our universe, as listed in Figure 9. All the energies in our universe exist on a continuum known as the electromagnetic (EM) spectrum.

Figure 9 Shutterstock image.

The EM spectrum is organized by wavelength of the vibrations of the energies in it. This implies that the energies travel in wave-like undulations. This understanding is often known as the wave-particle duality: energy in our universe is bundled in discrete units called photons, and yet those photons move through space in waves. The waves of energy are similar to waves we see in a large body of water.

Not all waves are the same size or distance apart. The amount of energy in a collection of waves is a function of the distance between the waves and of the frequency with which those waves crash into the beach or an object. Distance between waves is called the *wavelength*, and in light energy, as with all EM radiation, distinction is made using wavelength. In the ocean, distance between waves is often a matter of meters, or maybe even miles; with energy like visible light, the distance is a matter of *nanometers*, which are *billionths of a meter*. In other words, all of the wavelengths of energy, which comprise visible light in our EM spectrum in our universe, could fit in a band that is less than the width of the average human hair (Figure 10).

Which should impress upon you that we even have the ability to perceive this incredibly narrow band of energies with our eyes. And yet we do. Humans perceive the miniscule differences in these energies so well that most people can discern blue–green from aquamarine, and perhaps even more

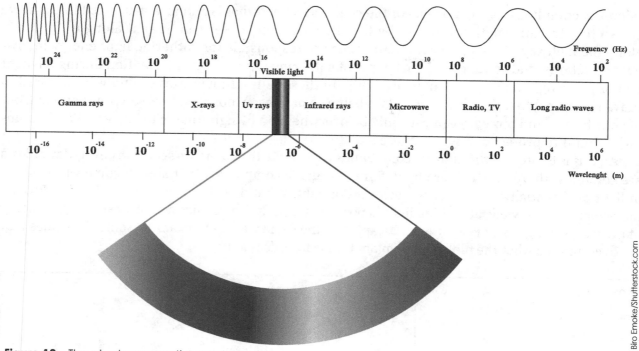

Figure 10 The electromagnetic spectrum.

subtle differences; consider the many different shades of blue found among the paint chips at a home store. How is this possible? The structure of the eye is basically a gel-filled sac with a clear lens in the front and a forest of specialized light-receptive cells along the inner rear wall of the orb (Figure 11).

Light enters the eye through the cornea, the clear covering of the eye. The amount of light allowed to pass into the eye is regulated by the iris, which is a ring of muscles that comes in colors like brown, blue, and green (and others). The pupil in the center of the eye looks black in the same way that a cave looks black from the outside entrance – there is no light in the inner part of the eye. The retina is a layer of blood-rich cells and nerves. The cells of the retina are composed of two main types: *rod cells* and *cone cells* (Figure 12).

Rod cells are long and thin; cone cells are squat and more pointy. Both are composed of the same basic parts: an upper section of layers of light-sensitive receptors filled with a protein called *rhodopsin*; a round

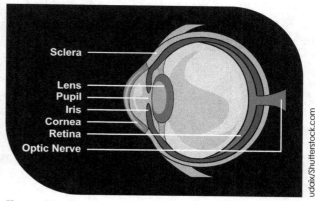

Figure 11 The structure of the eye.

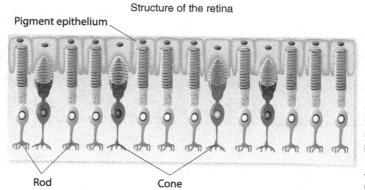

Figure 12 Rod and cone cells.

middle section housing the nucleus of the cell; and a 'foot' that is actually structured like a nerve cell so that the information received by the rhodopsin molecules can be sent along to the brain along the highway of the optic nerve. Rod cells are responsible for gathering light energy in low-light situations, and cone cells are in charge of gathering light energy information during daylight and brighter light situations. There are limits to these cells. Human eyes can't always see well in the dark – remember the phrase of a place being so dark as to 'not be able to see your hand in front of your face' – and our eyes can certainly be overwhelmed by light that is too bright. We involuntarily squint to protect our eyes in these situations.

All this is to present the kinds of compounds that create the colors we see. Many everyday objects are colored with *dyes,* which are by definition organic compounds that absorb some wavelengths of light preferentially over other wavelengths of light. What does it mean, to *absorb* light? From a molecular point of view, absorbing light is possible when dye molecules actually capture the energy of the waves of light and keep it from escaping in the same way. Let's examine some molecules that are dyes and see what the molecular explanation is for this reality.

Phenol Red	
Orange GGN – E111	
Yellow 2G – E107	

Green S – E142	
RS – E130	
Indigo carmine – E132	
Peonidin (Violet)	

What do you notice about these compounds? What do they have in common? How are they different? All are aromatic, many are phenols. Many contain nitrogen as well as carbon, hydrogen, and oxygen. Some have a measure of symmetry in all or parts of the molecule. Many have double bonds between atoms that are not in aromatic rings. These observations point to compounds of a similar class to phenols, but instead of oxygen atoms connected to the aromatic ring, these compounds also

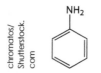

Figure 13 Aniline.

contain nitrogens connected directly to the aromatic ring(s). Compounds that contain groups of nitrogen and hydrogen atoms on an aromatic ring are called *aniline* functional groups (Figure 13).

Other structures are nitrogen atoms doubly bonded to each other; these are *azo* dyes.[15] All of the examples in the table above have one or more of these functional group arrangements: aromatic rings, phenol, aniline, and/or azo groups. You may also notice that several of the dyes contain $-SO_3$ groups, often with Na atoms. This is a sodium sulfite group, which usually brings stability to the dye molecule when it is mixed with other materials.

The dyes we use today have their start in nature. Nature's colors have been inspiration for thousands of years of human art and textiles. The earliest dyes were extracted from plants but suffered from fading in the sun and from washing. Modern dyes have found a way to overcome these limitation and become more *colorfast*. The first man-made dye is probably *picric acid*, which is a beautiful yellow. It was first synthesized in 1771, it only slowly became popular as a yellow dye for French army uniforms by the middle of the 19th century.[16] However, the tendency of this dye to be explosive quickly moved its use out of textiles and into munitions. The first commercially successful man-made dye is *mauveine*, created by accident by William Henry Perkin in 1856. Then a teenager in London's Royal College of Chemistry, Perkin was attempting to synthesize the antimalarial drug *quinine*, which was in great demand by British nationals stationed in malaria-plagued British colonies in India and Asia. Using coal tar as a starting material seemed like a good idea according to his professor, probably since there was so much of it lying around London since gas-lighting had come to the region. Gas street lamps and gas lamps in London homes were powered by the flammable gases released during high-temperature heating of coal. The problem was what was left after all the flammable gas was given off: coal tar. Certainly, Joseph Lister found use for it in extracting carbolic acid from it. But the coal tar contained much more than phenol. It contained many complex, aromatic compounds, many of which were dark and stained anything that the tar dropped onto. Coal tar extracts were beginning to be used in the infant area of microbiology – these extracts would sometimes adhere to bacteria in a way that would make them more visible under the microscope. This is where the connection between dyes and drugs begins.

William Henry Perkin attempted to synthesize quinine, but was unsuccessful except that he ended up with a black goo which, when dissolved in ethanol (the same compound that gives beer and wine their intoxicating effects) resulted in a beautiful purple solution of a shade that he had never seen before. As with the coal tar starting material, the purple dye was colorfast in all types of fabric. Nearly overnight Perkin's dye, now called *mauve,* was the rage among all the fashionable elite in London, and he became a young, wealthy entrepreneur (Figure 14).[17]

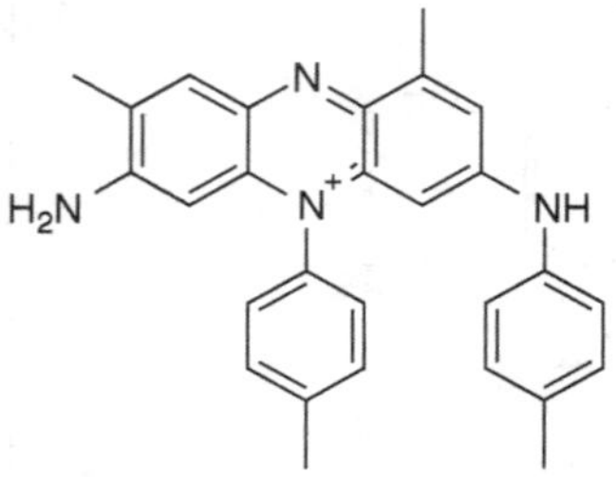

Examination of the compound as it is currently understood reveals the aromatic and aniline functional groups. These are most certainly the reason for the deep color and of the colorfastness of the molecule in many types of fabric. The commercial success of Perkin's serendipitous dye inspired others to do the same. Within a few years of Perkin's discovery,

Figure 14 Mauveine.

[15] *Encyclopædia Britannica Online*, s. v. "azo compound", accessed July 13, 2015, http://www.britannica.com/science/azo-compound.

[16] *Encyclopædia Britannica Online*, s. v. "picric acid", accessed July 13, 2015, http://www.britannica.com/science/picric-acid.

[17] LeCouteur, Penny, and Jay Burreson (2003). Napoleon's Buttons. P. 161. Jeremy P. Tarcher/Penguin Books: New York.

aniline dyes were being commercially produced by several European chemists and companies. Some of these, in Germany, made significant inroads in the understanding of coal-tar-derived aniline compounds as both dyes and as biologically active compounds. German scientists like Dr. Paul Erlich (coined the term *magic bullet* to describe a medicine that could affect only diseased cells and not healthy ones); Felix Hofmann, who's arthritic father was his inspiration for creating aspirin; and Gerhard Dogmak, whose experiments with Prontosil Red dye created the entire class of *sulfa* drugs, just to name a few.

Dyes Become Drugs (Figure 15)

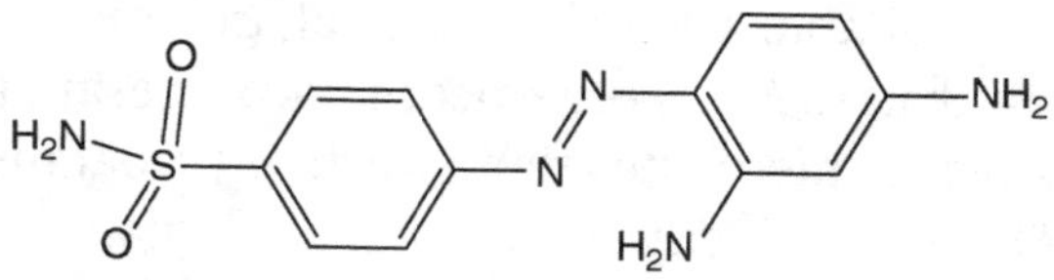

Courtesy of Author

Figure 15 Prontosil Red.

See how Prontosil Red has the aniline functional groups, an azo section, and is aromatic? These characteristics reveal it to be a decent dye. But it turns out that ingestion of this compound can also help cure certain types of bacterial infections because many biological systems will metabolize (break down) the dye molecule into *sulfanilamide* (Figure 16).

Why does this work? What does sulfanilamide have to do with killing bacteria? It turns out that many types of bacteria including those that causes meningitis, pneumonia, typhoid, and various staph infections are impeded by the presence of sulfanilamide in the person or animal suffering from the disease[18] (although only about 1% of all bacteria cause disease in humans).[19] Many of these bacteria have an internal mechanism that produces the vitamin folic acid (vitamin B-9) (Figure 17).

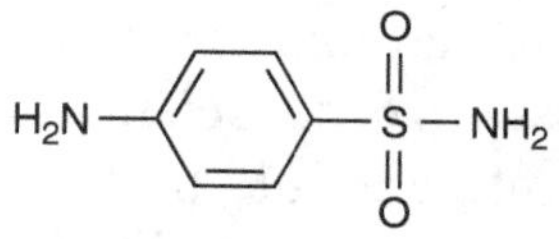

molekuul.be/Shutterstock.com

Figure 16 Sulfanilamide.

The central portion of folic acid has an aniline functional group with a C=O followed by another nitrogen (looking left to right across the molecule). The bacteria usually build this from materials present in the cells that they are infecting, but the bacteria are not discriminating consumers. When sulfanilamide molecules are shuttled to the infected cells, the bacteria only seem them as potential raw materials – and the sulfanilamide molecule looks *so much* like the central portion of the folic acid that a bacterium wants to make anyway (folic acid is an important material for the construction of new cell membranes as the bacteria reproduce during an infection in a host) that the bacteria simply inserts it lock, stock, and barrel into the folic acid synthesis, but of course, it doesn't work. Sulfanilamide is *not* exactly like the central

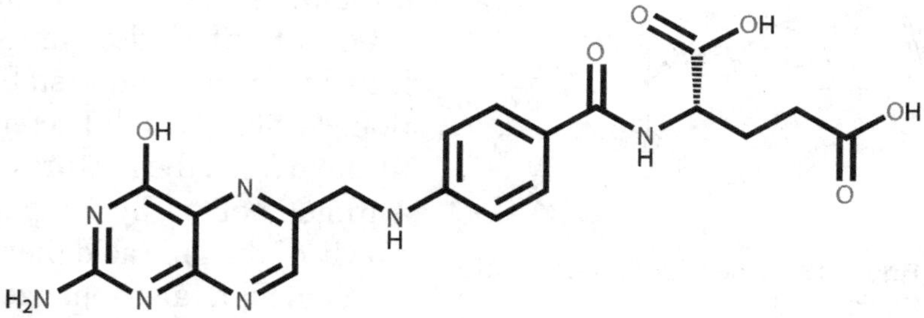

Leonid Andonov/Shutterstock.com

Figure 17 Folic acid.

---

[18] Rosenthal, Sanford M. Sulfanilamide Therapy of Bacterial Infections. *Science 10 February 1939: 89 (2302), 129–131. [DOI:10.1126/science.89.2302.129].*

[19] Ratini, Melinda (2015) "Bacterial and Viral Infections" *WebMD*. Available at http://www.webmd.com/a-to-z-guides/bacterial-and-viral-infections.

portion of folic acid, and so the folic acid is not made properly, and cell reproduction cannot proceed. The infection will not proceed and the host will recover as the sulfanilamide interrupts more and more folic acid syntheses.

Antimetabolites and Antibiotics

Sulfanilamide is an example of an *antimetabolite*. It is a man-made chemical that impedes the growth of bacteria and therefore promotes the end of infection. There are many derivatives of sulfanilamide made to treat specific bacterial infections, and in all of which the sulfanilamide molecules act as interruptions in folic acid synthesis.

There are other medicines available to treat bacterial infection. Not every kind of bacteria respond to the trickery of sulfa drugs. And some people are allergic to sulfa drugs. Most people think of penicillin when they consider medicines that effectively treat disease. Penicillin has a rather equally improbable story. Sir Alexander Fleming was a young British bacteriologist who was initially rather upset to find his lab was left a mess by his lab assistants while he was on vacation in Scotland. Returning to the lab in September 1928, he found some bacterial cultures of *Staphylococcus* bacteria colonies contaminated with mold identified as *Penicillin notatum*, which is a species of mold similar to the kind that grows on bread.[20] The interesting part was revealed by Fleming's microscope: where the mold spores were growing, the bacterial colonies were completely dead – in fact the bacterial cells that had come into contact with the mold were ruptured, and all the cellular contents were spilled out as if the bacterial cells had committed hari-kari.

Now *this* was really going to change everything. Fleming's accidental discovery alleviated his frustration with messy lab assistants, but the difficulty in growing the mold hampered plans to make this situation more than just an idea for a notebook. Several years pass, and the mantle of penicillin passes to other scientists better equipped to take the next steps. Research revealed that the mold was excreting a fluid that was causing the violent deaths of bacteria cells. In this fluid was a compound we now call *penicillin* (Figure 18).

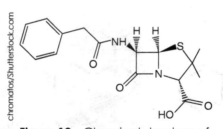

Figure 18 Chemical structure of Penicillin G.

Unlike sulfa drugs, penicillin and its derivatives are true *antibiotics:* they are compounds that directly kill bacteria. The action of antibiotics compared to that of antimetabolites is a little like death from a stab would compared to death by starvation. Antibiotics kill bacterial cells quickly and rather violently through *lysis*, the rupture of the bacterial cell wall or membrane. Antimetabolites, by comparison, kill bacteria by depriving them slowly of the folic acid they need to grow and reproduce.

Penicillin also changes the nature of disease treatment because of its wide effectiveness against many, many bacterial infection diseases. Diseases that cannot be affected by sulfa drugs may respond to penicillin treatment. Diseases like those caused by *Escherichia coli* (*E. coli*) bacteria are one example. When sulfa drugs and penicillin came onto the world stage during and after World War II, the earth's bacterial population had not been exposed to these compounds (at least at this magnitude) and as such was not prepared to resist the action of these drugs. Today, after decades of use (some might say overuse), strains of bacteria that were genetically susceptible to the action of sulfa and penicillin drugs have been greatly reduced, leaving behind strains of disease-causing bacteria that have developed genetic resistance – genetic mechanisms that prevent sulfa drugs or penicillin drugs from being

---

[20] Markle, Howard. (2013). "The Real Story Behind Penicillin." Available at http://www.pbs.org/newshour/rundown/the-real-story-behind-the-worlds-first-antibiotic/.

effective. In short, the bacteria have gotten smarter. To counteract this trend, the medical community has become more likely to prescribe rest for the first stages of disease, rather than a round of drugs, in order to allow time for a patient's own immune system to begin fighting the infection, and to discern whether an illness is viral or bacterial in nature. *Viral infections cannot be affected by antimetabolites or antibiotics.* Viral infections do not operate from as sophisticated a biological platform as do bacteria. As a result, treatments that are aimed at complex bacterial causes will not work for the simpler viral ones.

Viruses and bacteria may seem the same to the untrained person. And certainly neither is visible to the naked eye, and both can make a person or animal sick. But bacteria are true cells – they have a cell wall that encloses a true nucleus with DNA and other organelles that make the cell function. DNA are truly alive, in that they exhibit all the evidence of life: they breathe and eat, make wastes, grow, react to stimuli, and reproduce, all from mechanisms inside their own cell walls.[21] Viruses are much less complex. Most viruses only have a 'coat' called an *envelope* made of proteins, and no DNA. Instead, there is some a little DNA or some simple RNA (a derivative of DNA), which carries only the most basic information about how to get into a host cell and use the resources of the host cell to reproduce more viruses.[22] Viruses don't reproduce on their own, they don't grow – even this is enough to cast doubt on whether a virus is even alive (Figure 19).

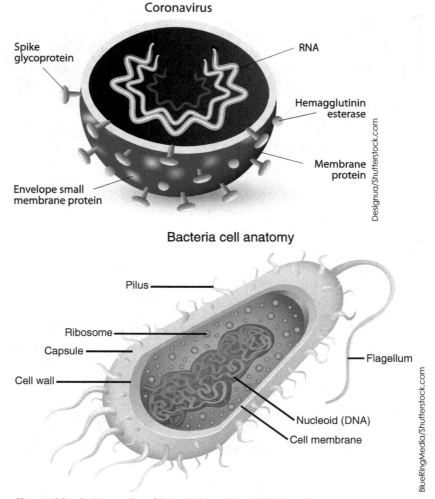

Figure 19   Schematic of bacteria cell and virus cell.

---

[21] "Life." Merriam-Webster.com. Accessed July 13, 2015. http://www.merriam-webster.com/dictionary/life.
[22] "Virus or Bacteria?" (n.d.) Available at http://www.microbeworld.org/what-is-a-microbe/virus-or-bacterium.

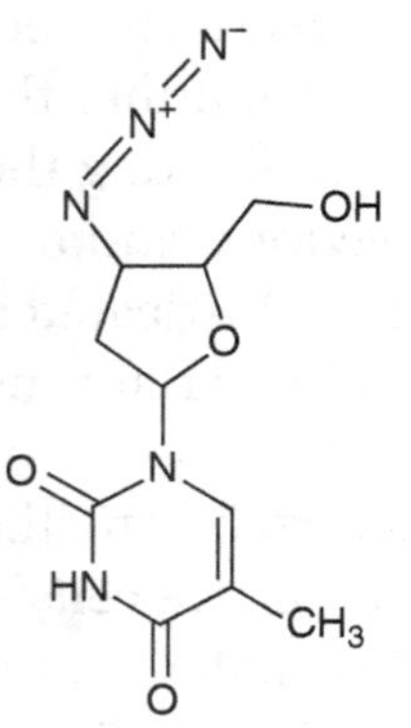

Figure 20 Retrovir, antiviral drug used for HIV/AIDS treatment.

Antiviral medications seek only to mitigate the effects of the viral infection. As science learns more about viral organisms, treatments are becoming more effective. Antiviral drugs for diseases like AIDS (Acquired Immune Deficiency Syndrome), caused by infection with the Human Immunodeficiency Virus (HIV), are more effective today at improving lifespan and quality of life (Figure 20).

It is obvious that this drug, Retrovir, which is one of the first antiviral drugs approved by the FDA (in 1987) is not a sulfa drug, nor is it a penicillin drug.

Other partner drugs to antibiotics and antimetabolites include tetracycline and cephalosporin (antibiotics), and metronidazole (an antimetabolite) among others. Three examples are shown here for reference. Look for molecular structures that are similar to sulfa or penicillin molecules (Figures 21, 22, and 23).

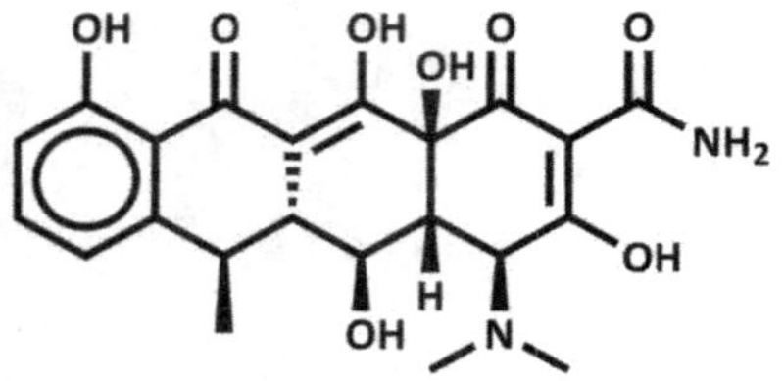

Figure 21 Doxycycline (tetracycline class)

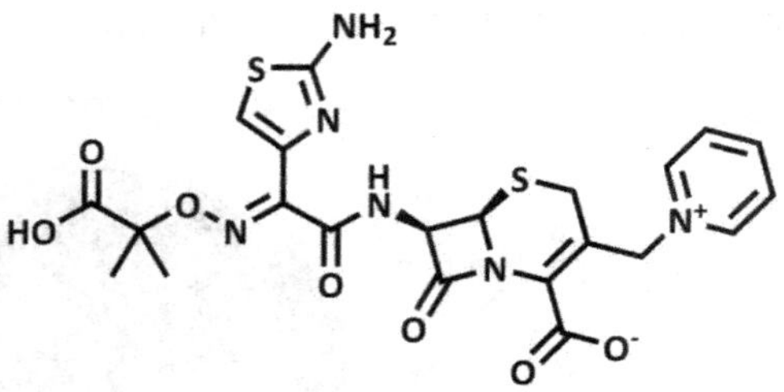

Figure 22 Ceftazidime (cephalosporin class)

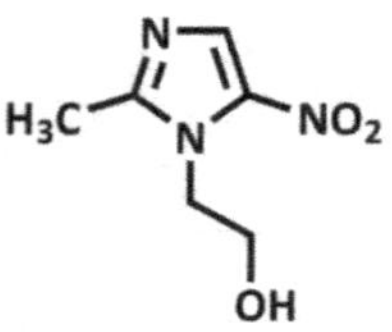

Figure 23 Metronidazole

STEROIDS AS DRUGS

When discussing analgesics, the classification NSAID was used. This applies to acetylsalicylic acid and ibuprofen. What does it mean to be a steroid, then?

The term *steroid* is a structural description of the molecular composition. All steroids have a basic structure of four rings, general known as ABCD rings (Figure 24).

Sometimes the six-sided rings may be aromatic, but not always. There will be side groups off of the rings are various places, but all steroids have the same central structure. One glance at the chemical structure should be enough to identify a steroid.

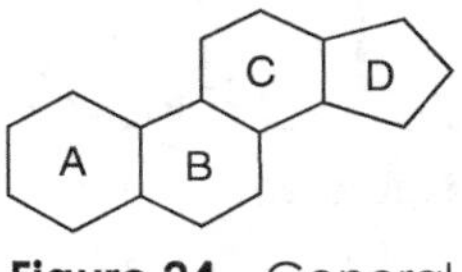

Figure 24 General structure of all steroids.

Steroids are often confused with *hormones*. The term hormone is a functional term. Hormones are compounds that may or may not be steroidal in structure, which are made by one organ or cell and are sent to tell another organ or cell to do something. Hormones are cellular messengers. In order to identify a molecule as a hormone, there has to be more information available about where the molecule is made and what the effect is of that molecule on other parts of the body.

Many steroids have anti-inflammatory effects on the body. One of the most common is hydrocortisone, also known as cortisol (Figure 25).

Cortisol is naturally made in the body to regulate sugar breakdown, maintain blood pressure, and in response to stresses of a variety of types: dieting, injury, illness, and exercise. Medicinally, cortisol is administered to reduce inflammation, to reduce allergic response, and to mitigate asthmatic response.[23] Given these uses, cortisol represents major health benefits and therefore is a drug in great demand. Obtaining cortisol naturally is really a process of tricking the body to either make more on its own or prevent the body from breaking down existing cortisol and is mainly a dietary approach.[24] But the amount of cortisol gained this way is relatively minimal. What about when there is a swollen arthritic knee or hip

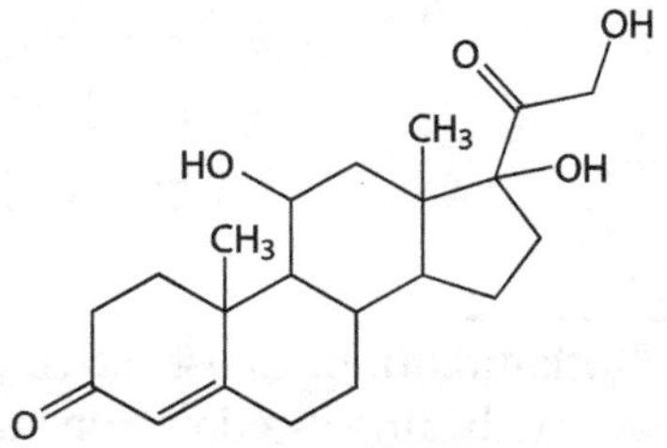

Figure 25 Cortisol.

---

[23] "Vitamins and Health Supplements Guide" (n.d.) Available at http://www.vitamins-supplements.org/hormones/cortisol.php.

[24] Francis, Lee (2013). "Natural Sources of Cortisol" *Livestrong.com.* August 16, 2013. Available at http://www.livestrong.com/article/332119-natural-sources-of-cortisol/.

that needs treatment right away? Or a major outbreak of hives that is causing intense itching? For greater amounts of therapeutic cortisol, a synthetic route to the molecule must be available. Until the 20th century, this was only a pipe dream.

Inflammation is caused by the influx of white blood cells and body fluids like pus that arrive when there is an infection or injury. Not all inflammation is bad – some research indicates that inflammation actually helps heal damaged tissue by bringing nutrients and other materials to the site of the damage so that cells can be repaired and replaced.[25] But when inflammation is causing undue pain or other discomfort, or if it is obstructing breathing, as in an anaphylactic response, then inflammation must be reduced. This is where anti-inflammatory drugs like steroids can be helpful.

Steroids have other uses as well. A common link is made between professional athletes and steroid use (or abuse). What is the drive to use steroids? Most athletes would say it is to improve athletic performance, or simply increase muscle size.[26] Some sources say it is to counteract pressure felt from the high expectations set from society, peers, fans, or coaches.[27] And what steroids are being used? Most 'performance-enhancing drugs' are derivatives of testosterone, the principal male sex hormone. This class of compounds is collectively known as *anabolic steroids*, which are known to increase muscle size by promoting muscle tissue growth (Figure 26).[28]

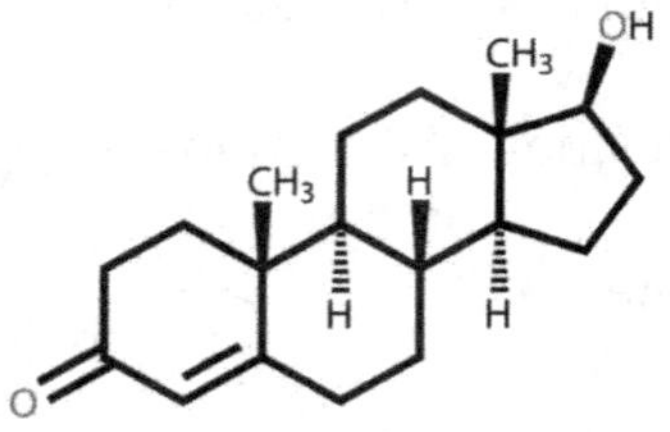

(See how the four rings identify the molecule as a steroid? But see also that the structure of testosterone is not the same as cortisol, though they are both classified as steroids.) Normally, testosterone is made in the male testes, along with the other androgen molecules, dihydrotestosterone and androstenedione. Women also have small amounts of testosterone made by the ovaries. For both sexes, testosterone is part of what builds and maintains muscle tone, as well as in the healthy function of the reproductive tissues. Higher levels of testosterone result in increased amounts of body hair, increased sexual drive, and a tendency toward baldness.[29] Using testosterone or its derivatives therapeutically has historically centered around the rebuilding of muscle mass after injury or illness, or treating hormone imbalances in males during or after puberty.[30] Examples of anabolic steroids includes testosterone as well as others (Figure 27).

Figure 26 Testosterone.

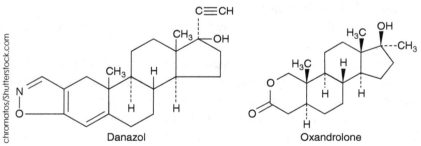

Danazol   Oxandrolone

Figure 27 Examples of anabolic steroid compounds.

---

[25] "Inflammation: Causes, Symptoms and Treatment." (2015). *Medicalnewstoday.com*. Available at http://www.medicalnewstoday.com/articles/248423.php.

[26] "Why do people abuse anabolic steroids?" (2006). http://www.drugabuse.gov/publications/research-reports/anabolic-steroid-abuse/why-do-people-abuse-anabolic-steroids.

[27] "Why Do Athletes Use Steroids?" (2014). http://www.why.do/why-do-athletes-use-steroids/.

[28] "Anabolic Steroid." Merriam-Webster.com. Accessed July 14, 2015. http://www.merriam-webster.com/dictionary/anabolic steroid.

[29] "Normal Testosterone and Estrogen Levels in Women."

[30] "Anabolic Steroids" *Medline Plus*. Available at www.nlm.nih.gov/medlineplus/anabolicsteroids.html.

No discussion of steroids and hormones would be complete without including the estrogens. Many people think that estrogen is a hormone or steroid itself, but this is incorrect. Estrogen is the name for the group of female sex hormones that has several members of which the main three are the steroids estradiol, estriol, and estrone. A companion steroid in women is progesterone (Figures 28–31).

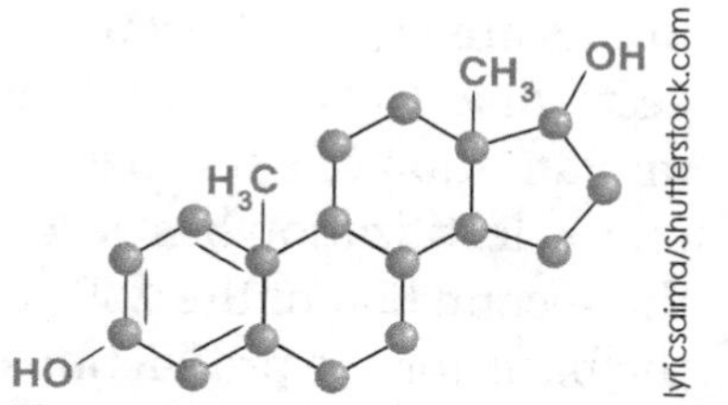

Figure 28   Estradiol.

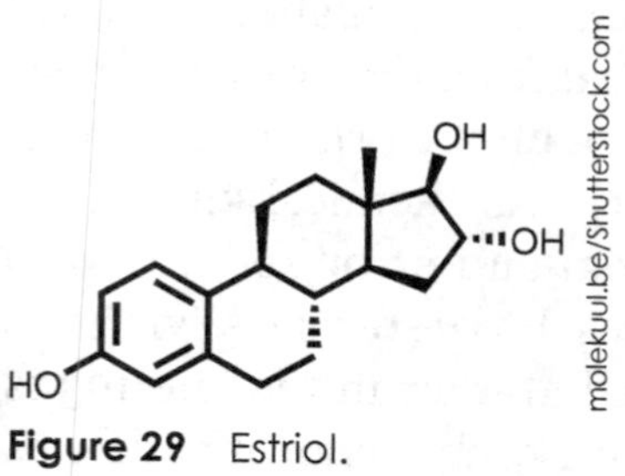

Figure 29   Estriol.

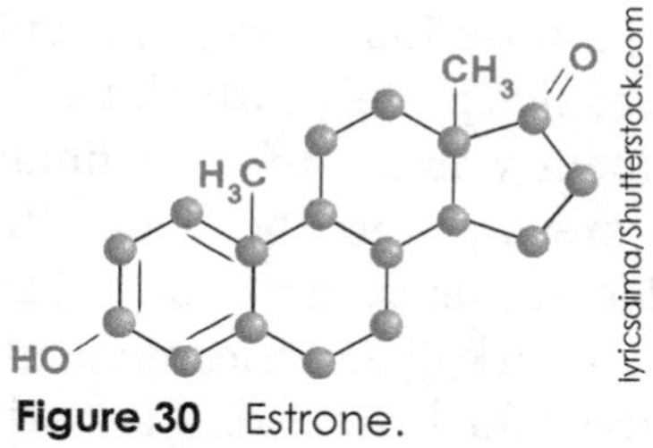

Figure 30   Estrone.

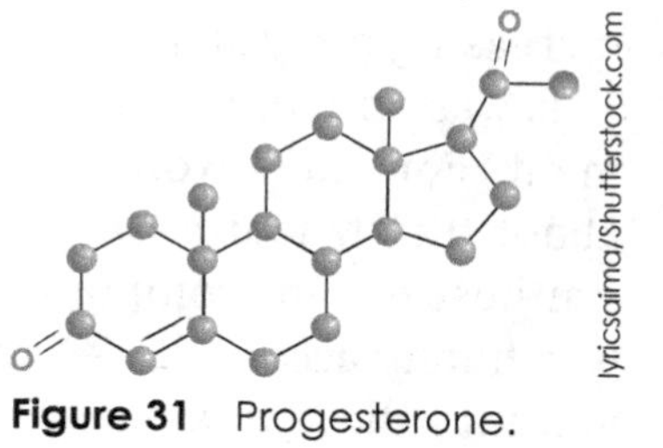

Figure 31   Progesterone.

In companionship with the molecule progesterone, as well as the follicle-stimulating hormone and luteinizing hormone, the estrogens regulate the menstrual cycle. In healthy women of childbearing age (classically understood to be best between ages 20 and 35),[31] this is a chemical symphony designed to prepare the uterus for possible implantation of a fertilized egg, a release of a matured egg from an ovary, and the dissolution of the unfertilized egg and the majority of the contents of the uterus at the end of the cycle if fertilization of the egg does not occur (Figure 32).

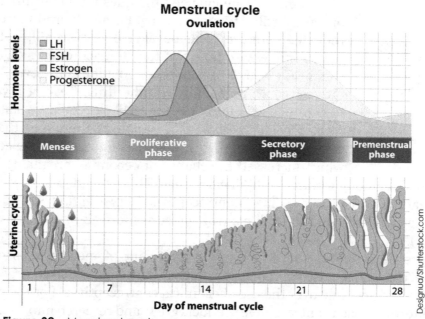

Figure 32   Menstrual cycle.

One application of the knowledge of male hormones has been the development of man-made anabolic steroids. Science has not ignored the female hormones either, but the biggest development

---

[31] "Best age for childbearing remains 20–35 – Delaying risks heartbreak, say experts" (2005). 9/16/2005. Available at http://www.medicalnewstoday.com/releases/30737.php.

in this area has not been a synthetic or derived version of an estrogen molecule, but instead of progesterone. In women, the main role of progesterone are 1) to prepare the lining of the uterus for implantation of a fertilized egg, if it shows up and 2) to prevent a second ovulation from occurring once a fertilization has occurred. This second action is the one that has garnered the most interest in the second half of the 20th century because the control of ovulation, which progesterone provides, is the basis for the production of the female birth control pill.

Research into the role of progesterone began in the early decades of the 20th century, mostly centered around fertility issues. Treatments for infertility caused by irregular ovulation included increasing progesterone levels through injection of progesterone extracted from the ovaries of pigs. This was an inefficient and labor-intensive process, which naturally drove up the price of the treatment. For many women, this treatment was out of their financial reach. As chemical knowledge about the structures of the steroids grew, so did knowledge through trial-and-error about how to synthesize many compounds in the lab, rather than rely on natural sources. Progesterone's synthesis was finally accomplished through the work of American chemist Russell Marker in the late 1930s and into the 1940s, using plant sources for his starting material. His work, now known widely as the Marker Degradation, resulted in a fall in the cost of progesterone fertility therapies due to an increase in the amount of product available.

Problems still remained. Progesterone is a liquid at room temperature and must be administered by injection. This specialized delivery is going to leave women out who cannot get access to the proper equipment for injections, as well as women who just can't stand needles. A more portable, accessible, stable form of the chemical was needed. This appeared within a few years of the Marker Degradation in the form of the compound norethindrone, created initially in 1951 by chemists who had studied Marker's work (Figure 33).

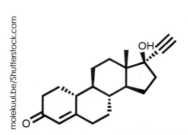

Figure 33 Norethindrone.

Norethindrone changed the question about the use of progesterone as a fertility treatment drug. Because norethindrone was a solid at room temperature, could be taken orally as a pill, and could withstand the caustic environment of the stomach during digestion to remain chemically active to suppress ovulation, now every woman with an interest could potentially use the drug. Testing revealed that it not only helped with infertility issues like irregular ovulation (suppressing ovulation chemically for a period of time, and then stopping that treatment sometimes initiated spontaneous ovulation for women, which increased the chance of becoming pregnant) but also reduced unwanted pregnancies among women of childbearing age. This expanded the use of norethindrone from a fertility treatment drug to also being a contraceptive. *Contraception* is the prevention of conception and pregnancy (compared to birth control, which is more about dealing with a pregnancy once it has started).

The FDA approved norethindrone in 1957 for treatment of menstrual disorders, including infertility. By 1960, demand resulted in expansion of the purpose to include contraceptive use. Social conflict erupted at the same time over the use of chemical contraceptives. For the next two decades, the use of norethindrone-based contraceptive products grew, and research into short- and long-term effects also identified the problems with these chemicals. Early versions of the contraceptive pill had levels of norethindrone that proved to be harmful (e.g., increased risk of blood clots, heart attack, stroke, and depression), and in 1988, lower-dose pills were marketed, which proved as effective as the original formulation without the same level of risk.[32]

---

[32] Nikolchev, Alexandra (2010) "A brief history of the birth control pill" *pbs.org*. May 7, 2010. Available at www.pbs.org/wnet/need-to-know/health/a-brief-history-of-the-birth-control-pill/480/.

Sociologists have been measuring the results of the societal and global impact of the pill. Some see the pill as the door for women to careers, sexual freedom, and reproductive control.[33] Others measure the pill as the straw, which broke the American family.[34] Who knew four little rings could do so much?

---

[33] Cohen, Sandy (2005) "Birth control pills helped empower women, changed the world." July 17 2005, *Copley News Service*. Available at http://religiousconsultation.org/News_Tracker/birth_control_pills_helped_ empower_women_changed_world.htm.

[34] Schlafly, Phyllis (2014). "Who Killed the American Family?" *wnd.com*. Sept. 22, 2014. Available at http:// www.wnd.com/2014/09/who-killed-the-american-family-2/.